Corrosion Damaged
Concrete

CIRIA, the Construction Industry Research and Information Association, is an independent non-profit-distributing body which initiates and manages research and information projects on behalf of its members. CIRIA projects relate to all aspects of design, construction, management, and performance of buildings and civil engineering works. Details of other CIRIA publications, and membership subscription rates, are available from CIRIA at the address below.

This book is an outcome of a CIRIA research project on concrete deterioration and repair. This work is continuing, guided by a Steering Group responsible to CIRIA's Advisory Committee for Building and Structural Design.

CIRIA's research manager for building and structural design is Eric Dore.

CIRIA
6 Storey's Gate
Westminster
London SW1P 3AU
Tel. 01–222–8891
Fax. 01–222 1708

Corrosion Damaged Concrete
assessment and repair

Peter Pullar-Strecker, MA, CEng, FICE

Construction
Industry
Research and
Information
Association

Butterworths
London Boston Singapore
Sydney Toronto Wellington

First published, 1987
  Reprinted, 1988

© CIRIA, 1987

British Library Cataloguing in Publication Data

Corrosion damaged concrete: assessment and repair
  1. Reinforced concrete——Repairing
  I. Pullar-Strecker, Peter   II. Construction
  Industry Research and Information
  Association
  624.1′8341   TA683

  ISBN 0-408-02556-5

Library of Congress Cataloging in Publication Data

Pullar-Strecker, Peter.
  Corrosion damaged concrete.

  Bibliography: p.
  Includes index.
  1. Reinforced concrete construction—Maintenance
and repair.   2. Concrete—Corrosion.   I. Construction
Industry Research and Information Association.
II. Title.
TA683.P85   1987   620.1′37   87–22577

  ISBN 0-408-02556-5

Photoset by Latimer Trend & Company Ltd, Plymouth
Printed and bound in Great Britain by Anchor Brendon Ltd, Tiptree, Essex

Foreword

Reinforced concrete is the most versatile and potentially one of the most durable materials that a designer can choose for almost any type of building or structure. Good durability can be achieved, but it does demand careful attention to detailed design, and painstaking supervision and workmanship during construction.

Because the required high standards are not always achieved, a small proportion of reinforced concrete structures is affected by durability problems which lead to the need for early repair.

This Guide has been written to help those who have to deal with the commonest of concrete's durability problems, the damage that results if reinforcement starts to rust.

The Guide is intended for the owner, for his professional advisers, and for their builders or contractors. Its purpose is to explain what has gone wrong and how the damage can be put right, so that informed decisions can be made and expert help can be sought when it is needed.

The Guide cannot convert the general engineering practitioner into a repair specialist nor can it provide detailed instructions for a repair contractor. Rather, the Guide provides the means of understanding what is happening so that errors will not be made through ignorance and inappropriate work will not be commissioned as a result of commercial pressure.

It has been written so that it can be understood by the non-specialist but at the same time help to increase the knowledge of those who are already well informed. A bibliography of further material, carefully chosen from the huge and growing literature on this subject, is included for those who want guidance on where to dig deeper into any of the subjects outlined in this Guide.

The Guide is not intended to show any geographical bias and its recommendations apply to a wide range of climates. The experience which was drawn on in writing this Guide has come from local sources in the UK, from the highway bridge repair programme in the USA and from the Arabian Gulf region where the hot salty climate is especially hostile to reinforced concrete.

Acknowledgements

The Guide was prepared in-house by CIRIA staff working under the guidance of a group of experts who were chosen to provide a wide spread of current experience of all aspects of concrete deterioration and repair.

The first Chairman of the Group was Mr Roger Hayward of Sir William Halcrow and Partners. He led the group with great skill and infectious enthusiasm until a few days before his death in July 1985 from a bravely fought illness. The Group remember him with the greatest respect and affection.

Constitution of the group at the time of publication

Keith Brook BSC CENG FICE, FIHT
  Chairman — Wimpey Laboratories Ltd

Phil Bamforth BSC CENG MICE — Taywood Engineering Ltd

Derrick Beckett BSC MPHIL DIC — Thames Polytechnic (formerly Sir Frederick Snow & Partners)

Bob Berry — Sealocrete (UK) Ltd
  Vice Chairman FerFa and Chairman of FerFa Working Party on Concrete Repair

Roy Bishop BSC PHD FICORRST — Consultant (recently retired Transport and Road Research Laboratory)

Bill French BSC PHD FGS — Queen Mary College

Richard Higgins BSC PHD ARCS DIC — International Paint plc

Brian Jones BSC CENG MISTRUCTE — BarFab Reinforcements

Doug Irvine BSC(ENG) CENG FICE — Tarmac Construction Ltd

Ted Kay BSC PHD CENG MICE — Travers Morgan & Partners

Tim Lees BSC CCHEM FRCS MICT — Cement & Concrete Association

Brian Long BENG CENG MICE — Consultant to Balvac Whitley Moran Ltd

Roger McAnoy BSC CENG MICE — Taywood Engineering Ltd

Chris Page MA PHD CENG MIN MICORRST — University of Aston

Ken Pithouse BACHEM DMS — Raychem Ltd

Derrick Pollock BSC PHD CENG MICE — Sir William Halcrow & Partners Ltd

Peter Robery BSC PHD CENG MICE — Taywood Engineering Ltd

Laurie Tabor RPRI — Fosroc Ltd

Graham Tilly BSC(ENG) PHD ACGI — Transport and Road Research Laboratory

John Tory BA — Department of the Environment

Ken Treadaway MSC CCHEM FRIC — Building Research Establishment

Jonathan Wood BSC PHD CENG MICE FIAGRE — Mott Hay & Anderson, Special Services Division

Former members of the group

John Whiteley BMET CENG MIM
David Thompson BSC PHD
Norman Jones MMS MBIM

Reinforcement Steel Services
Transport and Road Research Laboratory
Eurotunnel

Contributors of advice, photographs and other help

Laurence Collis BSC FICE MICT FGS
Nigel Wilkins BSC
Paul Langford MSC MICORRST
Avonguard Ltd
Building Research Establishment
Colebrand
Raychem

Messrs Sandberg
Harwell Corrosion Consultants
Taywood Engineering Ltd
(Photograph page 21)
(Photograph page 28 & Flowchart page 13)
(Photograph page 34)
(Photograph Section 2.12 page 76)

Contributors to the cost of the project

The UK Department of the Environment and CIRIA were major contributors. Other organisation contributing were:

Abu Dhabi National Oil Company (ADNOC)
Abu Dhabi Department of Public Works
BP International Ltd
DAFCO Concrete Products Industries, UAE
Dubai Municipality
Fairclough Civil Engineering Ltd
International Paint plc
Mott Hay & Anderson

Netlon Limited
Scott Wilson Kirkpatrick and Partners
Sir Alexander Gibb & Partners
Sir Frederick Snow & Partners
Travers Morgan & Partners and
   R Travers Morgan (Oman) Ltd
Wimpey Laboratories Ltd

CIRIA in-house staff engaged on the project

Project Director, Peter Pullar-Strecker MA CENG FICE, Deputy Director (now Pullar-Strecker Consultancy Ltd).

Technical Manager, Concrete Durability Programme, Mike Walker JP CENG MICE FIHT (now Concrete Society).

Research Assistant, Repair Database Project, Helen Emms, BSC (until August 1986). Darrell Leek MSC FGS (from September 1986).

Secretary to the Group, Deirdre Whitfeld.

The guide was compiled and illustrated, except where otherwise acknowledged, by Peter Pullar-Strecker.

The review of research and recommendations regarding chloride associated reinforcement corrosion in the UK and USA was prepared by Darrell Leek and Mike Walker.

Technical Editor: David Garner, CIRIA.

Contents

1 Introduction

1.1 Dealing with distress

If a reinforced concrete structure shows signs of cracking, spalling or rust staining, or any other sign that might indicate distress, the first task is to find out how serious it is, what has caused it, and how it can be put right.

There are many possible causes of distress; they include movement of the foundations, structural overloading, accidental damage, sulphate attack and alkali aggregate reaction, as well as the rusting of the reinforcement which is the subject of this guide. Any of them can need immediate action to secure the safety of the structure, but more often what is going wrong is progressing relatively slowly. The task is then not so much one of taking emergency action as one of trying to understand what is happening, trying to foresee how it will develop and putting in hand a plan to deal with it effectively and economically.

1.2 Why reinforcement rusts

When reinforcement rusts, the deterioration is usually slow, but it is always progressive. Effective action depends absolutely on having a proper understanding of how rusting is caused and how it can be controlled.

Since the beginning of reinforced concrete, the reinforcement needed to withstand tensile stresses or control shrinkage cracking has taken the form of bare steel bars, rods or wire which in theory are protected from rusting by the alkalinity of the concrete that surrounds them. In practice too, this protection is usually effective for a very long time, but it is possible for the protection to be reduced permanently either as the result of the penetration into concrete of acid gases which are present in the air (most often carbon dioxide, but in areas with industrial pollution, also sulphur dioxide), or by the penetration into the concrete of chlorides from seawater or from road salt used to thaw ice and snow. Calcium chloride, which was widely used in the past to accelerate the hardening of concrete, has a similar effect. The Appendix gives a brief history of the use of calcium chloride in concrete.

The ability of gases or salt to penetrate into concrete depends largely on the permeability of the concrete; low quality concrete can be quite permeable, but good quality concrete is relatively impermeable. Differences in the permeability and thickness of the concrete cover over the reinforcement largely explain the great time differences that can occur before the onset of reinforcement rusting in different structures which are apparently exposed to the same environmental conditions.

Although non-expansive corrosion in concrete is possible (see below), when reinforcement rusts the rust normally occupies a much larger volume than the steel and eventually exerts enough force on the concrete to cause cracks which slowly grow as the reinforcement continues to rust. Eventually all the concrete covering the reinforcement will be forced off by the growing rust and the reinforcement will be left not only without any remaining protection against rusting, but also without any bond to transfer the tensile stresses which it may carry to the concrete.

Structural failure will eventually follow, though the process leading to it is a slow one which normally gives years of warning, except sometimes in the case of prestressed concrete.

Note on post-tensioned concrete

Prestressed concrete uses the same principle of reinforcement to carry the tensile stresses, but there are some important differences. The first is that the reinforcement is deliberately stressed in tension before any load is put on the structure, and the other is that the post-tensioned prestressing may be anchored to the concrete only at its ends, being free to move within the concrete over much of its length. This difference means that rusting may be able to take place without any sign of it being visible on the concrete surface and structural failure can occur suddenly without warning.

The mechanisms of reinforcement corrosion have been the subject of intensive study and research during recent years and there are now many published papers which describe them in detail. Very briefly, the situation is as follows.

Steel in concrete can be in four possible conditions:

'passive', where it is protected from corrosion by a film of oxide which is stable in the environment of uncontaminated alkaline cement paste;

'generally corroding', where the steel rusts uniformly over its surface usually because alkalinity has been reduced as a result of carbonation, the rate of corrosion depending largely on moisture content and temperature;

'pitting', where the steel corrodes rapidly at a number of separate points as a result of chloride contamination at the boundary between the steel and the cement paste;

'unprotected (but cathodically restrained)', where the steel is in saturated concrete that does not contain enough oxygen to passivate it, but corrosion is *usually* restrained by lack of a cathodic current because the whole system lacks the oxygen needed to drive the cathodic (non-corroding) end of the reaction. However under conditions where only a small part of a single piece of concrete is saturated, corrosion can be rapid and lead to loss of steel without the formation of the familiar expansive red rust.

Severe pitting corrosion
in saturated concrete

In each case, corrosion is an electrolytic process which needs moisture in the concrete to provide a sufficiently conductive electrolyte (corrosion virtually stops at relative humidities of less than 50% in salt-free concrete and a somewhat lower figure in chloride-contaminated concrete), and oxygen at the non-corroding points (cathodes) to sustain the reaction which provides current for the corroding points (anodes). Like many processes which depend on chemical reactions, corrosion is accelerated as the temperature increases.

1.3 Cracking as a sign of reinforcement rusting

Apart from rust staining (except of course, rust staining caused by iron-staining aggregates), the earliest visible sign that reinforcement is rusting seriously is usually a hair-line crack in the concrete surface lying directly above the reinforcement and running parallel with it. Such a crack means that the expanding rust has grown enough to split the concrete.

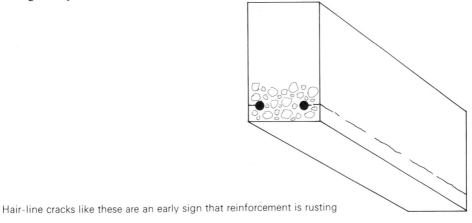

Hair-line cracks like these are an early sign that reinforcement is rusting

Because concrete is weak in tension and not flexible, very little rust is needed to crack it, and the reinforcement at this stage may look almost rust-free if the concrete is chipped off to reveal it.

But the fact that the concrete has cracked parallel with the reinforcement nearly always means it is rusting quickly enough to cause major trouble, and further damage becomes apparent after a time that depends on the conditions inside the concrete close to the reinforcement.

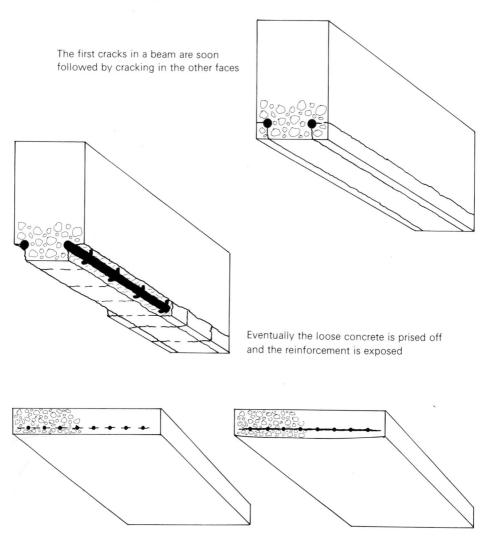

The first cracks in a beam are soon followed by cracking in the other faces

Eventually the loose concrete is prised off and the reinforcement is exposed

Cracks can form within the thickness of the concrete, reaching out from one bar to the next

The sagging of the concrete soffit has been exaggerated in the sketch. It would not be visible, but if the cover is tapped with a hammer the hollow sound will be unmistakable

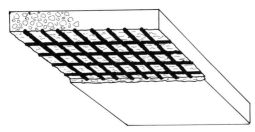

It is surprising how long a slab will remain supported in this way before it suddenly gives way

Not all cracks result from rusting, and whether hair-line cracks in reinforced concrete do indeed indicate that deterioration has occurred and will progress depends on the cause of the cracking and the position of the cracks in relation to the reinforcement.

The significance of different types of cracking is discussed in Section 2.2.4, but cracks at right angles to the main reinforcement (transverse cracks) do not normally indicate that the reinforcement is rusting. They are caused either by restrained contraction of the concrete soon after it has hardened (usually as a result of thermal contraction or shrinkage) or by stress from loads on the structure.

If these cracks remain fine (say less than 0.3 mm wide) they are also very unlikely to cause the reinforcement to rust unless they also coincide with reinforcement which runs parallel with them.

Cracks, even if fine, which lie directly over and run parallel with reinforcement, almost always result from the formation of rust and indicate that deterioration will progress.

The cracks in this precast concrete railway fence run parallel with the reinforcement and have been caused by the growth of the rust. In this case the concrete failed to protect the steel from rusting for two reasons: the cover was of poor quality and very thin, so it carbonated quickly down to the level of reinforcement, and the concrete contained calcium chloride admixture to accelerate hardening and increase the productivity in the precasting works.

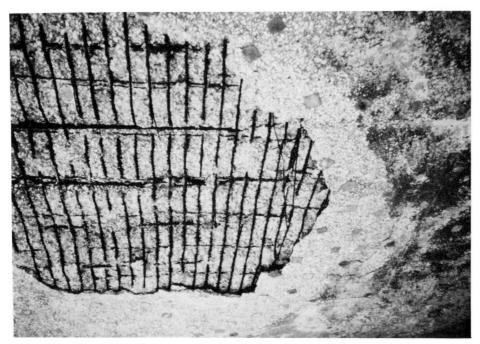

This badly delaminated concrete soffit of a covered walkway in Bahrain leaves no doubt that the poor quality concrete can have given very little protection to the reinforcement. The mortar spacers which are visible on the right of the picture are in much better condition than the bulk of the concrete. Even at the earliest signs of delamination or cracking, the extent of deterioration shown in this photograph could have been predicted with certainty from an examination of the concrete.

1.4 The importance of concrete quality

Provided the thickness of the concrete cover over the reinforcement is adequate, knowing the quality of the concrete protecting the reinforcement can be of some help in assessing whether early cracking is an indication that serious deterioration will follow. Quite a lot can be learnt about the general quality of the concrete by inspection with a practised eye, supplemented by listening with a practised ear when the concrete is tapped with a hammer. The presence of contamination or carbonation, however, cannot be discovered without carrying out some simple chemical tests.

1.5 The influence of the environment

The aggressiveness of the environment is a very important factor to consider when examining concrete that shows signs of possible distress. In a sheltered indoor environment, minor cracking or some minor defect in concrete quality (but not chloride contamination) may not lead to serious trouble, whereas in severe exposure like sea-splash or alternate wetting and drying, especially in hot arid climates, only the highest quality of undamaged concrete can be expected to give satisfactory performance in the long run.

When concrete is permanently saturated, it is normally protected from corrosion by lack of oxygen. (See Section 1.2)

Taken together with evidence from a study of cracking and examination of the soundness of the concrete, the general conditions of exposure and aggression of the environment can be a useful indicator of what rate of deterioration can be expected.

De-icing salt used on road bridges can create a local artificially aggressive environment and this has led to damage costing several billion dollars in the USA. In northern Europe, similar but less extensive damage has been caused to bridge substructures by de-icing salt washed from the decks by the rain.

Environments which are generally, rather than locally aggressive, are easier to identify. They include coastal areas where structures can be exposed to splash and spray from the sea, or wetted and dried by the tide; and industrial plants which emit chemical pollution.

Signs of concrete distress in aggressive environments like these usually mean that serious deterioration will follow. The question of what action should be taken is considered next.

This balcony in Bahrain, although built from inferior concrete and contaminated with salt-laden dust and sand carried by the wind, shows advanced failure through the rusting of the reinforcement only in the area which has been wetted by water from the air conditioning unit (the pipe to lead the water from the unit to the ground is a recent addition, added no doubt when the damage was already obvious). In time, when rusting has destroyed most of the reinforcement, this balcony will collapse and fall into the street below.

Another example of an environment which is aggressive only locally in part of a structure is this ring beam which incorporates a built-in rainwater channel in the upper face. This structure is also in the Arabian Gulf and the channel fills with salty sand blown from the coast. The dew and occasional rain have washed the salt into the concrete where it has become concentrated by evaporation and has destroyed the protection the concrete originally gave to the steel.

1.6 Options for repair

Reinforced concrete structures which show signs of deterioration should be surveyed and tested by specialists, who will be able to determine the causes of the deterioration and advise on the likely future life of the structure and the options for repair. In the light of this advice, the owner responsible for the structure will have to decide what action to take. Very broadly, the options are:

1. Do nothing other than carry out regular safety inspections. Allow the deterioration to run its course
2. Take action to prevent the deterioration from getting worse.
3. Carry out repairs to restore deteriorating parts of the structure to a satisfactory condition.
4. Demolish and rebuild all or part of the structure .

As well as the technical questions, there are strategic questions that have to be answered before an option can be chosen: What are the owner's future needs and intentions? What are his resources and wish to invest in the structure? Is he obliged to keep the structure serviceable or at least in a safe condition?

None of the options can be considered sensibly without some knowledge of what degree of restoration is possible. Broadly, this depends on the cause of the damage, the quality of the concrete, the extent of any contamination by chlorides and how far the deterioration has progressed.

1.6.1 Accidental loading

If a reinforced concrete structure has been damaged by accidental overloading of some sort (including excessive early shrinkage or thermal stresses) and if the concrete is of good quality and uncontaminated by chlorides, there is no reason why the structure should not give entirely satisfactory service once the damage has been repaired.

Cracks which have formed as a result of accidental overload will tend to be very fine when the load has been removed and often they need no treatment at all. Cracks more than 0.3 mm wide may need to be sealed against ingress of moisture, acid gases or salt by injecting resin into them under pressure, if they are not expected to move again.

If further movement is expected at the position of a crack which needs to be sealed, it must be widened and made into a joint which can be sealed with an elastic sealing compound in exactly the same way as movement joints which are made intentionally.

The treatment of cracks in sound concrete is an almost everyday engineering procedure and has stood the test of time in many different situations. It is considered in more detail in Section 4.1.3. Accidental damage to reinforced concrete through impact or other temporary overload (excepting fire damage) can be repaired to restore the structure to its original service life, provided that repairs are carried out before the reinforcement has started to rust and before the damaged surfaces are affected by carbonation or contamination.

1.6.2 Faulty construction

Faulty construction is one of the most common causes of early deterioration.

Common construction faults include inadequate compaction and failure to position the reinforcement so that it has adequate concrete cover. Under almost any exposure conditions these faults will eventually reduce the service life of the structure as a result of reinforcement rusting after the concrete has become carbonated.

If action is taken quickly, before the reinforcement has started to rust, it is possible to prevent or greatly reduce the chance or extent of damage from these causes. Voids in under-compacted concrete can be injected with cement grout or resin, extra concrete cover can be added by concrete spraying or screeding techniques, or the impermeability of the cover can be increased by applying a suitable resin (polymer) based coating.

1.6.3 Faulty materials

Concrete which is contaminated with a significant level of chloride, introduced either with the original mix materials or as a result of chloride penetration from a salty environment, must be cut away and replaced with uncontaminated concrete wherever it is close to the reinforcement.

The significance of different levels of chloride contamination is discussed in Section 2.5, and methods of cutting away and replacing concrete cover are considered in Section 4.

1.6.4 Maintenance of preventive systems

Apart from the special cases of accidental damage or damage caused by faulty construction, it is difficult to prevent further deterioration of already deteriorating reinforced concrete, except by replacing it.

The owner responsible for a deteriorating reinforced concrete structure must accept that he may be committed to continuing periodic maintenance or repair to keep the structure in a serviceable condition. The cost and frequency of future maintenance or repair will depend on the type and condition of the structure, on the cause of its deterioration and on the type and quality of the initial repair work or preventive system.

Speedily carrying out repairs and providing extra protection for structures which are known to be deteriorating at a significant rate may be a way of avoiding much larger repair costs at a later date if the protection is provided soon enough, but it is doubtful whether such repairs or extra protection will themselves be maintenance free. For example, structures which are protected with surface coatings (Section 5) will need to be recoated from time to time, and cathodic protection (Section 6) needs continuing monitoring and control.

1.6.5 Repair strategy

Repairing reinforced concrete by what could be called engineering methods has become widespread only during recent years, and there is hardly any experience of how well such repairs last in practice.

Consequently, repair strategies tend to one of two extremes: low-cost patch repairs done on a 'wait and see' basis with the acceptance that further repair will probably be needed from time to time, and high-cost fundamental repairs carried out with the hope that a permanent cure will be achieved.

Both strategies have their place, the choice often depending on the ease of access, the availability of maintenance facilities, and availability of funding.

In future years, the relative economics of the alternative systems will become clearer as evidence of performance is built up.

1.6.6 Replacing structural elements

An alternative to repair or continuing preventive maintenance is the replacement of deteriorating parts or even of the whole structure.

Replacement is almost always much more expensive than repair. It usually creates operational difficulties as well as formidable structural problems in providing temporary support and then in ensuring that the replacement member carries its correct share of the structural load. Nevertheless the replacement of members is a well proven and reliable method of repairing a structure and sometimes is the only safe or economical course of action in the long term.

1.6.7 Demolition and reconstruction

A reinforced concrete structure which has deteriorated very badly may be beyond economical repair. If it is still needed, then there is no alternative to replacing it with a new structure.

No doubt the lessons learnt from the failure would incline the owner to be cautious about rebuilding in reinforced concrete. There are alternatives for most structural forms, but the probability is that reinforced concrete construction of good design and built to a high specification would be at least as durable as any of the economically feasible alternatives. Reinforced concrete structures that have failed after only a comparatively short period of service have usually done so because of inadequacies in design, quality of materials, construction, or all three. This need not apply to the replacement structure.

The importance of designing and constructing to achieve durability in reinforced concrete was underestimated until comparatively recently. Mixes were designed and concrete was produced almost entirely on the basis of structural strength; designers sometimes paid little attention to the need for detailing to avoid attack from aggressive environments; codes and specifications were often optimistic in their requirements where they concerned durability; and constructors were unaware of the degree of care that was needed to avoid the causes of early deterioration.

The owners of reinforced concrete structures can expect long and economical service from reinforced concrete structures which are rebuilt taking full account of today's knowledge of how to achieve durability.

In the sections of this Guide that follow, great emphasis is placed on the need for the highest standards in repair work, so that the life of the repairs and of the structure that is repaired are the longest possible in the circumstances that apply.

1.7 The need for assessment: a summary

Reinforcement in concrete is usually protected from rusting by the alkalinity of the concrete. Sometimes this protection is lost because of carbonation or contamination or both. When this happens the first signs of longitudinal cracking are usually quickly followed by spalling and eventually by loss of cover. Repair options range from continuously maintaining preventive treatments to demolishing and replacing the structure completely. Between these extremes are various methods of repair and protection that can maintain the structure in a serviceable condition. With the present state of knowledge repairs cannot yet be regarded as permanent if rusting has started, but they can extend the useful life of a structure. In some cases it may also be possible to retard or prevent deterioration without carrying out repairs. Knowledge of the causes of deterioration has increased greatly during recent years. Repair or reconstruction based on a thorough understanding of these causes, careful specification and conscientious workmanship can be expected to produce structures with a much greater life.

(See references 1 to 12 for further information on topics in this section).

Corrosion damage in a UK jetty.

2 Assessment of structural condition

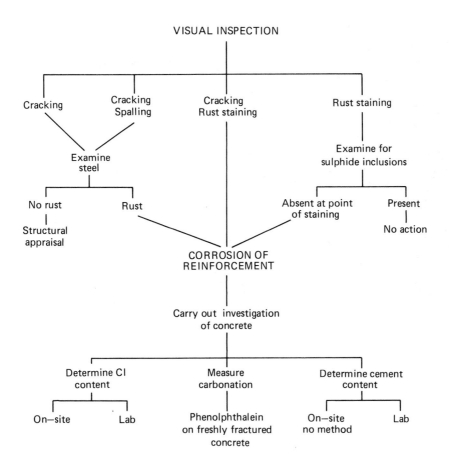

Flowchart for inspection of corroded steel in concrete.

The previous section explained how some reinforced concrete structures can be affected by rusting reinforcement, and it described some of the signs that may mean that progressive deterioration is taking place from this cause.

If visual inspection of a structure suggests that such a problem may be affecting it, the next step is to make a careful examination of the structure and carry out tests which will positively identify the cause and extent of the trouble, and allow some prediction to be made about the future life of the structure.

This section describes the methods of assessing the condition of structures affected by rusting reinforcement.

2.1 Inspection and testing

Inspecting and testing structures which are thought to be deteriorating should be done in two stages: first, an initial survey using simple methods to establish whether there is a need for repair, and second, a thorough survey and testing of representative parts of the structure to allow a repair scheme to be designed and its cost to be estimated.

Even the simplest methods benefit greatly from experience and a knowledge of what to look and listen for: the more specialised methods absolutely demand the use of skilled experienced operators on site to carry out tests and take samples, in the laboratory to analyse materials and in the office to interpret the results in the light of the design of the structure and the implications of its present condition for its future performance.

The normal methods suitable for an initial survey are:

1. Visual inspection and soundings of parts that can be reached from the ground, using binoculars to inspect and telephoto photography to record details that are out of reach. Hand magnifiers and scales can be used to estimate crack widths. Hammer tapping (or chain-drag on slabs) can be used to locate hollowness or delamination (internal splitting and separation).
2. Testing parts that can be reached from the ground for thickness of cover, position of reinforcement, and concrete strength, all by non-destructive methods using simple instruments.
3. By breaking off small pieces of concrete, testing freshly exposed concrete surfaces with indicator solution to determine the depth of carbonation.
4. Measuring the chloride content of pieces of concrete broken from the structure, or dust samples obtained by drilling holes into it.

Extending any of these simple tests to parts that cannot be reached from the ground or from the structure itself needs access equipment and increases the cost considerably. The more specialised tests described next would normally be done only after the results of an initial survey had shown that repair was likely to be needed. At this stage it can be assumed that access equipment will usually also be needed.

5. Mapping the half-cell potential of the concrete surface relative to the reinforcement using a high-impedance millivoltmeter and a half-cell.
6. Locating areas of delamination, voids or weak concrete by measuring ultrasonic pulse-velocity through the concrete.
7. Measuring the electrical resistivity of the concrete.
8. Measuring the surface water absorption of the concrete in situ by ISAT or other test.
9. Testing core samples cut from the concrete for strength, permeability, contamination, composition and density.
10. Measuring changes in crack width with a strain gauge.
11. Measuring the deflection of structural members under known applied loads.

Basic equipment for visual inspection and sounding in addition to normal safety equipment

1. Drawings showing how the structure was designed. Failing that, a location sketch and sketches of any important details must be made on site.
2. Camera, with optional 50 mm and telephoto lenses (about 200–300 mm) for general views and inaccessible details.
3. Small steel rule with 0.5 mm divisions for measuring crack widths, or optical crack width gauge.
4. Hand lens, preferably 8 × to 10 × magnification 'linen tester' for use with steel rule, or specialised instruments for crack examinations such as the Pentax monocular scope or the Wexham microscope.
5. Folding rule for measuring one-handed and reaching inaccessible places.
6. 15 m tape for general dimension and location measurements. A plumb line and spirit level are also useful.
7. 6 × to 10 × magnification binoculars for inspecting inaccessible places.
8. Medium hammer (about 50 g) for sounding concrete for hollowness, and
9. Sharp 12 mm cold chisel for chipping concrete (for example for carbonation test) and for estimating concrete hardness.
10. Kneeling mat to encourage close inspection of floors and slabs. A lightweight extending ladder or stepladder is needed for close inspection of most soffits.
11. Tape recorder for making 'no-hands' notes, preferably a small voice-activated recorder that can be hung around the neck.
12. Hand brush for cleaning dirty surfaces.
13. Wax crayon or self adhesive labels for marking reference points on the concrete surface so that they can be identified in photographs.
14. Power drill, 20 mm or larger masonry bits, rubber cup or funnel, collecting bags and adhesive tape for collecting concrete 'dust samples'.
15. Compressed gas can (of the type used for cleaning electronic connectors) or puffer to blow dust from test holes.
16. Phenolphthalein spray for locating carbonated concrete.
17. Blowlamp for drying wet concrete surfaces.

Specialised equipment for more detailed inspecting and testing

1. Covermeter ('pachometer' in USA) for locating reinforcement and estimating cover thickness.
2. Schmidt rebound hammer for estimating concrete strength.
3. High impedance low range voltmeter and half cell for mapping electrode potentials. (See Section 2.12).
4. Resistivity meter for measuring concrete electrical resistance.
5. Ultrasonic pulse-velocity apparatus for locating voids and delamination.
6. Demountable mechanical strain gauge and adhesive disks for measuring crack movement, or tell-tales.
7. Access equipment for reaching inaccessible places.
8. Borescope for examining internal surfaces of core holes cut in the concrete.
9. Strain measuring equipment for live load tests.
10. Other specialised measuring equipment such as instrumented delamination detectors, thermography equipment or ground-effect radar may be useful on large sites.

Damage from carbonation.

Construction joints are a common cause of trouble.

2.2 Inspection of cracking

Cracking can be an important sign that reinforcement is rusting. Because the significance of each type of crack is different, it is important to identify and record details of cracks during the initial survey.

2.2.1 Types of cracks

Seven types of cracks need to be distinguished:
1. Longitudinal cracks which formed after the concrete hardened;
2. Transverse cracks which formed after the concrete hardened;
3. Shear cracks which formed after the concrete hardened;
4. Plastic-shrinkage cracks which formed in the unhardened concrete;
5. Plastic-settlement cracks which formed in the unhardened concrete;
6. Map cracks which formed in the concrete surface over an extended period of time;
7. Surface crazing which formed in the concrete surface over an extended period of time.

2.2.1.1 Longitudinal cracks

Cracks that run directly over reinforcing bars in such a position that they could not have been caused by shrinkage, plastic settlement, or thermal contraction, must have been caused by build up of rust forming on the reinforcement. These cracks are a symptom of deterioration which will eventually lead to spalling and complete loss of cover. They cannot be treated without removing and renewing the concrete cover. The repair of these cracks is discussed in Section 4.1.5.

2.2.1.2 Transverse cracks which formed after the concrete hardened

Transverse cracks which form in the concrete after it has hardened are caused by shrinkage, thermal contraction, or structural loading.

The width of this type of crack is usually controlled by the reinforcement, either the main reinforcement which carries stresses from the working load, or the secondary reinforcement, usually fixed at right angles to the main reinforcement for the specific purpose of controlling shrinkage and thermal cracking.

Where there is no secondary reinforcement, these cracks will only be transverse to the main reinforcement and will be harmless unless they are very wide or the environment is exceptionally aggressive.

Where reinforcement runs in two directions at right angles, cracks that are transverse to one bar will be parallel with another. When this happens, it is quite likely that the parallel crack will coincide with the bar because reinforcement of the larger sizes tends to act as a crack inducer. The repair of this type of crack is discussed in Section 4.1.3.

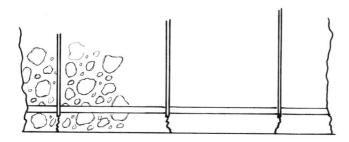

Cracks which are transverse to the main steel may lie over stirrups and eventually cause rusting. Conversely . . .

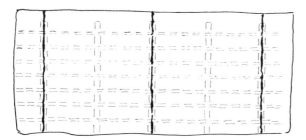

cracks which are transverse to the secondary reinforcement may coincide with the main bars and cause serious rusting if they are not sealed soon after they have appeared.

2.2.1.3 Shear cracks

Shear cracks are caused by structural loading or movement and are not the result of reinforcement rusting. They may however cause rusting if they are left untreated. The treatment of this type of crack is discussed in Section 4.1.3.

2.2.1.4 Plastic shrinkage cracks

These cracks form during construction on a concrete surface, such as a floor or road slab, if rapid evaporation causes a large moisture loss from the surface.

These cracks are harmless unless the concrete surface will be exposed to salt or contaminated dust, in which case they provide a reservoir for contamination at a level close to the reinforcement. The treatment of this type of crack is discussed in Section 4.1.1.

2.2.1.5 Plastic settlement cracks

These cracks form during construction in concrete which settles and bleeds excessively in the formwork. They tend to form longitudinally over the reinforcement, and are a common cause of serious rusting. The importance and treatment of settlement cracks is discussed in Section 4.1.2.

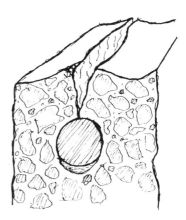

If a concrete mix is very workable, the solids can settle and allow the water to bleed to the top, especially if the lift is deep. Settlement cracks can form where the reinforcement has supported the aggregate and stopped it from sinking. Water which collects under the reinforcement displaces the cement grout and leaves the reinforcement unprotected.

2.2.1.6 Map cracks

Map cracking is caused by alkali-aggregate reaction and does not result from rusting of the reinforcement.

2.2.1.7 Surface crazing

Surface crazing results from the shrinkage of the concrete surface during or after hardening. It does not result from reinforcement corrosion.

2.2.2 Visibility of cracks

2.2.2.1 Cracks 0.05 mm to 0.1 mm wide

On a clean concrete surface which has been wet and is drying off, it is possible to locate very fine cracks by the visible lines and patches of dampness which follow the cracks and linger for a while after the rest of the surface has become dry. Under these conditions cracks less than 0.05 mm wide can be identified quite easily. Thoroughly spraying a concrete surface with water and allowing it to dry off is a good way of creating this condition artificially, though it may take half an hour or more for the cracks to show up. Conversely, drying the surface of wet concrete with a blowlamp produces the same effect. In all these cases, the line of the crack is seen most clearly at a distance from the concrete, because at closer examination the direction given by patches of dampness is less noticeable.

Strong *slightly* oblique lighting also helps to show up very fine cracks. A torch can be useful, or if it is sunny a magnifying mirror (shaving mirror) can help to direct the light especially if the general natural lighting is shaded.

Cracks that are so fine as to be visible only under such conditions may be innocuous, but if they are parallel with the reinforcement they can indicate quite an advanced state of deterioration.

2.2.2.2 Cracks more than 0.1 mm wide

Cracks become easily visible when they are about 0.1 mm wide and at about 0.5 mm wide the separation between the two edges of a crack can be seen by the unaided eye.

2.2.3 Crack width

With experience, the width of a crack can be estimated by eye but it is more reliable to use a crack width gauge, or an optical instrument specially designed for estimating crack width.

Usually there is no need for great precision in these measurements because it is necessary only to know whether the crack is less than about 0.3 mm wide, greater than 0.5 mm wide.

Where inspections are repeated periodically, it is useful to record whether crack widths or lengths have increased.

Crack width	Visibility
Less than 0.05 mm (2/1000 in) to 0.1 mm.	Noticeable only when drying out. Once noticed, visible in strong light.
0.1 mm (4/1000 in) to 0.5 mm (20/1000 in)	Noticeable with unaided eye. Separation of surfaces may just be visible.
0.5 mm (20/1000 in) or more	Both edges of crack visible as distinct interruption in concrete surface.

Simple optical estimation of crack width.

2.2.4 The significance of cracking

Longitudinal cracks expose long lengths of reinforcement to possible sources of contamination and they must always be regarded as potentially dangerous, even if they are not already the result of existing rusting, as is often the case.

Transverse cracks may or may not be serious, depending on their width, location, and the exposure of the structure: their treatment is discussed in Section 4.1.3.

2.3 Sounding the concrete surface

Reinforcement rusting that is causing the cover to separate from the body of the concrete can be detected by tapping the concrete surface with a hammer. Any heavy steel object (a large bolt, spanner, or length of reinforcing rod) will do, but a hammer is more convenient and gives more consistent results. For normal cover thicknesses a medium hammer is best, but for thick cover (75 mm or more) a small club hammer is needed to make the concrete move enough to sound hollow.

With practice, the sound of delaminated concrete can be identified accurately enough to allow the boundaries of the affected area to be sketched on the surface of the concrete or on a drawing of the structure.

If parts of the concrete cover have split internally from the body of the concrete, serious rusting is likely to be taking place and visible damage will follow, even if the surface looks intact at the time of inspection.

Floor slabs and road decks can be sounded by dragging a suitable object over them ('chain drag' is a term used in USA) or by tapping with a long steel rod. This saves a lot of bending down and consequently gets the survey done more thoroughly. On large sites, instrumented delamination devices or temperature-sensing photography carried out while the temperature is changing can help to pinpoint areas of delamination.

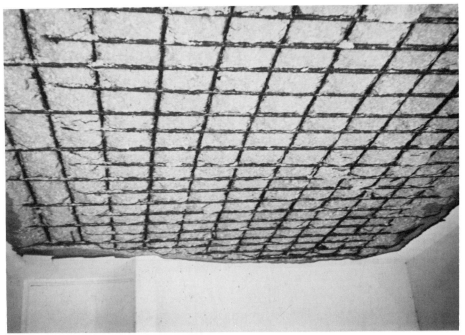

Soffits can also be tapped with a long steel rod, but beware of the danger that the concrete may fall if it is tapped too hard.

2.4 Testing for carbonation

Because the alkalinity of the concrete normally protects reinforcement from corrosion, testing the concrete to discover where it has lost its alkalinity through carbonation is a very useful guide to where and when the reinforcement may rust.

But note that concrete which is contaminated with chlorides may fail to protect the reinforcement even if no alkalinity has been lost through carbonation.

A very simple test of alkalinity is to spray the concrete surface with a chemical indicator solution which changes colour according to the alkalinity of the concrete. A solution of phenolphthalein in dilute alcohol is usually used because it has a very strong pink colour that is easily visible on any kind of concrete which has retained its alkalinity but is colourless on concrete which is no longer alkaline enough to protect the steel from rusting.

The change in colour of phenolphthalein takes place as pH increases from 8.2 to 10.0. This conveniently corresponds with the pH needed to protect steel by passivation.

To test whether concrete has carbonated, the surface tested must be freshly broken and then sprayed with pH indicator solution. A hammer and sharp cold chisel can be used to remove part of the concrete surface for this test, or a piece can be broken out at a corner section with a hammer. To make a test which is not adjacent to an edge or corner, a piece of concrete can be broken out if a ring of holes is drilled round it. *In either case, be sure that this test will not reduce the cover significantly at a vulnerable part of the structure and be ready to patch the damage with high impermeability proprietary repair mortar if there is any doubt.*

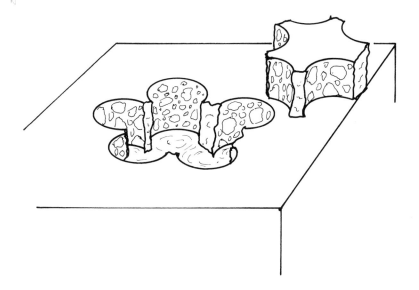

Note: It might be thought possible to estimate the depth of carbonation by drilling a hole into the concrete and testing the drill dust from various depths with phenolphthalein indicator. This method does not give reliable results because drilling tends to break up and expose the unhydrated elements of the cement (which are alkaline) and gives a falsely optimistic result.

The line indicating the edge of the carbonated layer is usually very irregular so it is best to record the maximum, minimum and apparent average depth of carbonation.

The outermost parts of the structure will be carbonated and will not be stained pink. As fresh concrete is exposed deeper into the structure, uncarbonated concrete will be reached. The boundary of the pink stain will clearly show how far carbonation has penetrated.

In good quality concrete which is highly impermeable only the very surface layer of a few millimetres thick will be carbonated, even in quite old structures. In very poor concrete, carbonation may have penetrated as far as the reinforcement. If it has, the reinforcement will have started to rust, though in a dry atmosphere (concrete inside a building for example) the rate of rusting may be so slow that it creates no problem.

In a moist atmosphere, or where concrete is alternately wetted and dried, reinforcement will start to rust as soon as the carbonation front reaches it, and in a salty environment the combination of chloride and lost alkalinity can cause very rapid rusting indeed.

Measuring the depth of carbonation into the concrete is a test which can give warning of rusting before serious damage has occurred. It may sometimes be possible to delay (or perhaps even to prevent) the carbonation from reaching the reinforcement by coating the concrete surface with a suitable material. (See Section 5).

Carbonation penetrates more quickly near corners—usually where the main reinforcement comes closest to the surface.

Carbon dioxide can penetrate into concrete where it has cracked, and along the reinforcement where it is locally debonded.

Carbonation also penetrates more deeply into the concrete where it is poorly compacted (e.g. honeycombed). It is important to test the depth of carbonation in such places.

Carbon dioxide enters the pores of the concrete as a gas and neutralises the alkalinity of the pore fluid. Carbonation is slow in saturated concrete because the pores are blocked with water, and also slow in very dry concrete because there is little pore fluid to react with. Carbonation takes place most quickly at relative humidities around the level that is required for greatest comfort; about 65%. Hence it tends to be fast inside occupied buildings.

Respiration of the pores caused by alternate wetting and drying or heating and cooling also speeds carbonation.

When the depth of carbonation and the cover over the reinforcement are known (See Section 2.6), repair or prevention can be planned on the basis of the rate at which carbonation appears to be progressing and the time at which the reinforcement will lose its protection from the alkalinity of the concrete.

Because both the depth of carbonation and the cover will vary, some parts of the reinforcement will be affected before others. The difference in time can be several years and this may make it economic to carry out partial repairs over an extended period.

2.5 Determining the chloride content

The chloride content of concrete is critical to the life of the reinforcement and quite small amounts of chloride can disrupt the oxide layer that should protect the steel from rusting. The risk that reinforcement will rust as result of chloride contamination depends on the concentration of the chloride, the alkalinity of the concrete, the amount of sulphate present, the type of cement and whether the chloride was present at the time of mixing or whether it penetrated the hardened concrete.

Once passivity has been lost, the rate of rusting will also be influenced by the moisture content and temperature of the concrete, and the availability of oxygen at different parts of the reinforcement.

These factors are complicated by their interaction, and they are not yet fully understood in spite of intensive research and many publications.

A broad guide based on currently accepted views is given overleaf.

Chloride determination in concrete is therefore a very important part of any survey to assess the condition of deteriorating concrete. Unfortunately there is no satisfactory spray-on indicator that reveals chlorides in the same way that phenolphthalein reveals carbonation, but simple analytical devices are available for use on site and give a useful indication of chloride levels even if they are not as accurate as standard laboratory analyses.

To determine the chloride content of concrete, samples have to be obtained, either by breaking off pieces of concrete or by drilling holes in the concrete and collecting the dust produced. Where salt has built up on the surface of the concrete, as the result of spray from the sea for example, the surface salt must be removed before the

Chloride by wt of cement	Condition of concrete adjacent to reinforcement	Corrosion risk
less than 0.4%	1. carbonated	high
	2. uncarbonated, made with cement containing less than 8% tri-calcium aluminate (C_3A)	moderate
	3. uncarbonated, made with cement containing 8% or more C_3A in total cementitious material. (*Note*: BS12 and ASTM C150 specify no LOWER limit for C_3A content).	low
0.4–1%	1. as above	high
	2. as above	high
	3. as above	moderate
more than 1.0%	all cases	high

Where sulphate is present at more than 4.0% by weight of cement, or where chloride has entered the concrete after it has hardened (rather than being incorporated at the time of mixing), the risk of corrosion is increased in all cases.

samples are taken. By collecting the dust from different depths separately, it is possible to determine how the chloride content changes with depth from the surface. This is important in finding out whether the chloride was present in the concrete when it was cast, or whether it entered later from the surroundings.

Obtaining samples by drilling is a quick and simple procedure which allows many

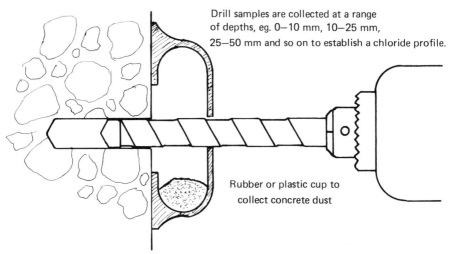

Drill samples are collected at a range of depths, eg. 0–10 mm, 10–25 mm, 25–50 mm and so on to establish a chloride profile.

Rubber or plastic cup to collect concrete dust

Drill samples are collected at a range of depths from several different holes to make up a representative sample for testing. The bit diameter should be at least 20mm.

samples to be taken so that representative results can be obtained in spite of the inevitable variation between individual samples.

The concrete samples are treated with acid to dissolve the cement, and the chloride content is determined by titration against silver nitrate. Chloride ion meters and rapid field test methods are also available, e.g. Quantab and Hach.

Experts tend to disagree over the accuracy of the different methods of analysis, but for many purposes that is unimportant because even approximate values are usually enough to establish whether or not chloride contamination is likely to be a cause of deterioration.

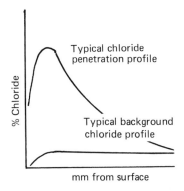

A typical profile for chloride which has penetrated the hardened concrete, and one for chloride which was present in the fresh concrete.

2.6 Estimating the thickness of cover

The significance of carbonation or of cracks found in the concrete surface depends on their position in relation to the reinforcement. Drawings of the structure will indicate where the reinforcement was intended to be and what sizes of bar were specified. Even if such drawings are available, it is best to check the actual position of reinforcement and particularly the depth of cover. Electromagnetic cover meters are helpful in skilled hands, though the information they give can be confusing if bars are closely bunched. The depth of bars is sometimes difficult to disentangle from the bar size because a small bar near the surface can give much the same reading as a larger bar at greater depth.

Covermeter readings can estimate the depth of the reinforcement to about + or − 5 mm but they should first be calibrated on site by using the meter to locate the reinforcement and then drilling to determine its depth.

2.7 Measuring crack movement

Cracks caused by rust growing round the reinforcement increase in width much too slowly for movement to be measurable even during a lengthy condition survey, but cracks caused by thermal or structural loading may still be responding quite quickly to changes in the load. It is important to know if they are, because in repair work live cracks must be treated as movement joints (which indeed they are, even if unintended) whereas dead cracks can be grouted or covered over unless they are the result of rust formation. (See Section 4.1.4)

Various forms of electrical strain gauge are available and some are suitable for use in a site survey, but the most practical and convenient way to detect and measure crack movement is with a mechanical strain gauge that can be held on gauge discs glued to the concrete surface.

Demountable strain gauge for measuring crack movement.

Cracks which are due to applied loads will move immediately the load is changed (traffic passing over a bridge, or heavy furniture or sandbags moved on a floor for example). Cracks due to thermal movement move while the temperature of the element is changing, for example early in the morning as the day becomes warmer, or when the sun reappears from behind a cloud.

Measurements are quick and easy to make, and measurements made three or four times throughout the day should establish whether a crack is live or not. A dummy measuring point here and there on an uncracked part of the concrete helps to identify whether movement is real or due to the operator's lack of technique.

2.8 Estimating in-situ strength

Concrete which has been well compacted and has a high cement content is likely to be high in strength and low in permeability. Because of this, strength tests can help to indicate whether the concrete is likely to be resistant to the penetration of chlorides and carbon dioxide. But there are exceptions: it is quite possible to make high-strength concrete which is fairly high in permeability (for example by compacting thoroughly at low cement contents and low water/cement ratios) and the converse is also possible, though less likely.

Do *not* assume that high-strength concrete will necessarily be more effective in preventing reinforcement rusting, but other things being equal, this is likely to be the case.

The way concrete resists a hammer blow or an attempt to cut it with a sharp cold chisel gives a broad indication of its strength to an experienced observer. Concrete which is easy to damage in this way is likely to have a very low strength – less than $15 \, N/mm^2$ – and will probably not protect the reinforcement against an aggressive environment unless the low strength is due solely to the use of a weak aggregate.

It is more difficult to estimate the strength of good quality concrete in a similar way, even with experience, but a simple non-destructive test can be made with a rebound hammer. This is a spring-loaded impacting device that incorporates a scale to measure the energy of the rebound following the impact. The instrument is sensitive to the moisture content, hardness and surface finish of the concrete (and whether or not a piece of aggregate is hit), but the test is very quickly done and the average of 20 or so impacts does give a quick means of estimating the concrete strength and, what may be more important, how the strength varies in different parts of the structure. A recording version of the hammer is available, but calling the results to a pocket tape-recorder is just as good and allows comments and location information to be recorded at the same time so that it can subsequently be marked on the concrete or a drawing.

Local areas of low strength in a structure probably mean that quality control was poor and indicate where to look more closely for incipient durability problems.

Many other non-destructive or semi-destructive strength tests have been invented, but they are less convenient to carry out and do not give much more information than the rebound hammer.

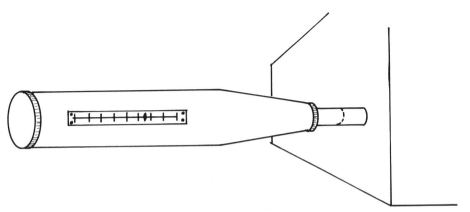

The Schmidt hammer is cocked by pushing it against the concrete surface like a grease gun. The impacting weight "fires" as soon as the plunger is fully home, and the rebound energy is indicated by the position the pointer returns to. Several tests can be made in less than a minute.

2.9 Taking core samples

The only way to get much more information about concrete strength is by cutting cores from the structure.

Cores of 100 mm diameter cut with diamond-edged coring cylinders allow a wide range of tests to be carried out, but cores as small as 50 mm diameter can be used even for strength testing. Small diameter cores can be cut dry with tungsten carbide tipped coring bits except in the hardest aggregates.

Core cutting is relatively expensive, but cores give a lot of information. They can be tested in compression to measure the concrete strength; they can be weighed to measure its density; sliced to measure its permeability; examined petrographically and analysed chemically to determine cement content and type, chloride content, water:-cement ratio, aggregate type and grading. And if a core is cut through the reinforcement, its condition can be examined too, though normally coring through reinforcement should be avoided because it can seriously weaken the structure.

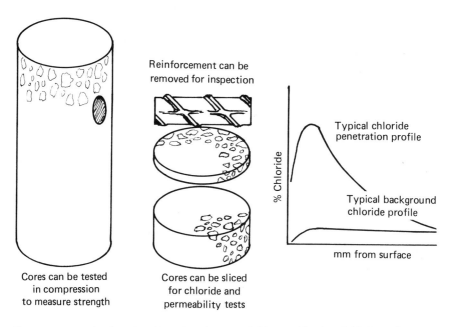

Reinforcement can be removed for inspection

Typical chloride penetration profile

Typical background chloride profile

% Chloride

mm from surface

Cores can be tested in compression to measure strength

Cores can be sliced for chloride and permeability tests

Cores are expensive to cut and test, but they can yield a wealth of useful information.

When cores are taken, it is important to inspect the walls of the core hole to confirm the positions of voids and breaks found in the core. This can be done with a borescope, or even a dentist's mirror.

If simple tests on a concrete structure indicate that there is doubt about its ability to perform adequately throughout its intended service life, coring the structure is going to be necessary anyway and expenditure on other forms of complicated test is probably wasteful.

Cutting even a small number of cores from a load bearing member can dangerously weaken it. The load-paths through the structure must be assessed and highly stressed areas or areas where there is no adequate alternative load path must not be weakened by coring.

2.10 Making ultrasonic pulse velocity (UPV) measurements

The speed of travel of an ultrasonic pulse through concrete depends on the density and elastic modulus of the concrete and thus gives an indirect measure of strength and perhaps durability. Voids and internal cracks can be found by this method, because the air gap greatly increases the transmission time.

The normal way of measuring UPV is to send pulses through the concrete from a transmitting probe on one side to a receiving probe on the other. This means that both faces have to be accessible, which is of course not always the case.

To estimate crack depth, both probes are applied on the same surface and the pulse transmission time depends on the length of the path round the bottom of the crack. The PUNDIT is a commercially available portable ultrasonic generator/timer specially developed for concrete testing.

UPV testing is certainly valuable in locating faults in concrete, but faults are not usually the cause of durability problems, though they may sometimes be the result, for example where rusting has caused internal rupturing or the total loss of the reinforcement.

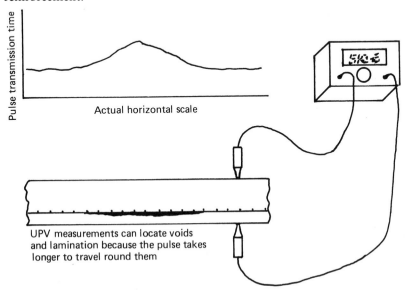

Actual horizontal scale

Pulse transmission time

UPV measurements can locate voids and lamination because the pulse takes longer to travel round them

Ultrasonic scanning which gives a picture according to the density of what is scanned is used in medicine but has not been applied to civil engineering.

2.11 Measuring water absorbtion and permeability

Measurement of the water absorbtion of the concrete surface by ISAT or a similar test method may be a way of determining how easy it is for contamination to enter the concrete and this may indicate whether deterioration through carbonation or salt penetration is likely.

Like resistivity (see Section 2.13), surface water absorbtion depends on the moisture content of the concrete at the time the measurement is made. It is very difficult to get meaningful measurements on the outside surfaces of concrete structures because the moisture content here cannot be standardised, but measurements on interior surfaces in buildings may be more consistent. Even then, the interpretation of the results in terms of future durability is very uncertain.

The permeability of concrete under standardised conditions can be measured in the laboratory on slices cut from concrete samples - for example from cores. Measured in this way, the results can give an indication of the danger that salt or carbon dioxide will penetrate as far as the reinforcement and put an end to the passive condition of the steel, and will give some indication of how freely oxygen can pass through the concrete to sustain the corrosive reactions (See Section 1.2).

2.12 Making electrode-potential surveys

Reinforcement which is in its ideal non-corroding state ('passive') is covered with a very thin layer of oxide which is continuously being maintained and repaired by the alkaline materials which surround it. Loss of alkalinity or the presence of certain contaminants (for example chlorides) destroys this film or prevents it from being repaired. (See Section 1.2)

The electrical potential of steel which is protected in this way is different from the electrical potential of steel which is in a condition where it can rust. Under favourable conditions, this fact can make it possible to use electrical measurement from the surface of the concrete to get an indication of how the risk that reinforcement might corrode varies in different parts of the structure.

Note that very dry concrete may have too high a resistivity to give an indication that the steel is not passive, and the test can then give unwarranted reassurance that all is well inside the concrete.

To make electrode potential measurements, electrical connections have to be made with the reinforcement and with concrete surrounding it. Electrical connection with the reinforcement is made simply by making a hole in the concrete to expose part of the reinforcement and connecting a wire to it. Electrical connection with the concrete has to be made by means of an electrolyte which wets both the concrete surface and a conductor to which another wire can be attached.

The electrolyte and the conductor are made up into a probe which can be held against the concrete surface. This probe is known as a half cell, the other half of the cell being the reinforcement and the concrete.

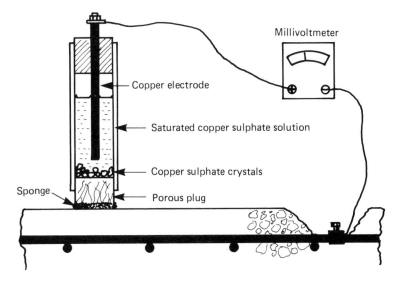

The essentials of a practical copper-copper sulphate half cell.

Together the two half cells form a complete electrical cell that generates a potential that can be measured with a high-impedance millivoltmeter (not less than 10 Mohms).

Copper with saturated copper sulphate is the most commonly used half-cell in concrete testing but many other combinations, for example silver with silver chloride, can be used. Each half-cell probe generates its own potential, the value depending on the materials used. The potential of the reinforcement relative to the pore fluid plus

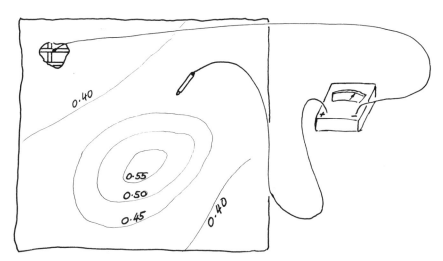

Lines sketched on the concrete join points of equal potential.

the potential of the probe gives the total potential difference which is measured with a high impedance millivoltmeter. It is usual to quote this total potential and the probe material when reporting potential measurements, e.g. '-0.42 V Cu/CuSO$_4$'.

It is sometimes claimed that electrode-potential measurements can be interpreted to discover whether or not reinforcement is likely to be rusting, but most experts now believe that this is not possible. The value of electrode-potential measurements is in comparing the state of the reinforcement in adjacent areas of concrete, for example in electrode-potential mapping.

To survey large areas, a number of half-cells may be mounted together on a suitable carriage. The 'Pathfinder' is a device of this type which incorporates a data processor and recorder to relate potentials to their position on the concrete surface.

2.13 Measuring electrical resistivity

Rusting of reinforcement is an electromechanical process which depends on the movement of electrically charged ions through the pore liquid in the concrete (See Section 1.2). The same movement of ions causes concrete to be electrically conductive, and measuring conductivity (or resistivity) gives a measure of how easily corrosion current can flow as a result of the potential differences caused by corrosion conditions.

The resistivity of concrete is related to the moisture content and quality. Resistivity is typically between 8000 and 12 000 ohm-cm but can be less than 5000 or greater than 15 000 ohm-cm.

The results of resistivity measurement are very sensitive to the conditions under which they are made, and they are also influenced by the proximity of the reinforcement.

Nevertheless, resistivity measurements can give information which helps the interpretation of electrode potential results made under the same conditions.

Resistivity in concrete is usually measured by the four-probe method. An electric current is passed between outer probes and the potential difference generated between the inner probes gives a measurement of resistivity.

Using an alternating current for resistance measurement prevents the probes from polarising and developing their own half-cell potentials with the pore fluid in the concrete.

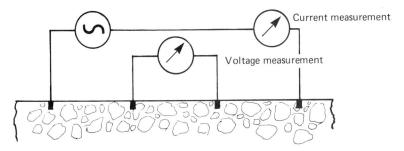

Resistivity measurements with four equally-spaced electrodes set in holes drilled in the concrete (Wenner method).

2.14 Summary of recommendations for assessment strategy

If there is reason to think that a reinforced concrete structure may be deteriorating because the reinforcement is rusting it should be inspected by someone who has experience in this field.

A simple inspection should show whether serious deterioration could be taking place. If this is thought possible, assessment of the structural condition and causes of the problem will be necessary.

The relatively simple and inexpensive tests that can be carried out on parts of the structure that can be reached without access equipment will show whether it is necessary to carry out further testing, and whether there is a need for access equipment to be installed.

If this further testing is done and shows that repair work will be needed, it may be best to consider immediately what repairs should be done so that arrangements can be made for them to be carried out while the access equipment is still in place. This not only saves time and money if repairs are going to be needed, but also allows the condition of the reinforcement to be confirmed by cutting away concrete to the level of the reinforcement here and there so that the full extent of the problem can be seen.

Working in this way can be difficult contractually because the extent of the work needed emerges only as it is being carried out, but there is no practical alternative. Usually, the problems are seen to increase as the work proceeds.

Commissioning inspection and repair work is considered in the next Section.

Further action will be necessary if any part of the structure is potentially unsafe, or if immediate repairs could prevent the deterioration from becoming worse, or if the serviceability of the structure (possibly including its appearance) is unacceptable. These are judgements that have to be made at the time and on the site, preferably on the advice of an independent person who is able to give a disinterested view based on an extensive previous experience and a good understanding of the causes of deterioration.

(See references 2, 8, and 11 to 24 for further information on topics in this section).

Staining caused by iron-staining aggregates, not reinforcement corrosion.

3 Specifying and commissioning inspection and repairs

The Introduction to this Guide sketched the general nature of the reinforcement corrosion problem and broadly what could and should be done to assess and tackle it. Other sections have explained how to identify damage and carry out repairs. This chapter tackles the question of how to commission inspection and repairs when they are needed.

3.1 Finding appropriate advice

Deterioration caused by reinforcement corrosion is commonplace only in certain specific types of structure: buildings made from precast components likely to contain calcium chloride, parts of highway bridges exposed to the run-off from de-icing salts, coastal structures exposed to splashing from seawater and structures built with insufficient or poor quality cover over the reinforcement. In other types of structure, damage from reinforcement corrosion is at present comparatively rare, though it can be expected to become more common as reinforced concrete structures age. This form of damage has been considered important only during the last 6 to 7 years in temperate climates, and for a little longer in hot coastal climates. The result is that knowledge of how to assess the extent and likely consequences of damage, and understanding of how best to repair it are not yet widespread (or complete) and have not passed into general practice in the construction industry. Expert advice is available, but the first decision has to be whether it is needed.

Reinforcement corrosion usually causes a type of apparently superficial damage which looks as if it can be repaired by the simplest means with the lowest technology and everyday equipment like a bucket of mortar and a trowel. The vast majority of repairs are probably carried out in exactly this way, and although they may not last for very long, re-repairing the repair is quick and cheap and must often be the most economical way of working. Provided that the damage is not endangering the stability of the structure, temporary repairs can continue in this way for many years, though it must be borne in mind that if a repair has to be repeated from time to time, the reinforcement is continuing to rust and will eventually be too weak to carry its share of the structural load.

Permanent repair, or repair of damage which is endangering the structure or people who come near it, needs expert understanding and skilled execution to be successful.

Some firms of civil or structural consulting engineers now have appropriate expertise either in-house or on call and can commission surveys by appropriate specialists to establish the causes of deterioration and advise on what action should be taken. If anything other the simplest repair work is needed, consultants can also design suitable repair schemes and commission contractors with the necessary skills and equipment to carry them out.

Commissions for consultants and contracts for repairers must be very flexible because it is usually impossible to foresee the full extent of the repairs that will be needed until access has been arranged and the work of uncovering the rusting reinforcement has begun.

Inevitably the contractor for repairs has to be sought from organisations experienced in this type of work, and he must be allowed some responsibility for the detailed design of repairs and the specification of materials, subject of course to the approval of the consultant.

3.2 Public relations

Repairs are usually carried out to buildings and structures which are occupied or in use and a good relationship between the occupiers and the repairers is essential to the smooth running of the contract. Security, noise, dust, access, storage of materials, hours of working and damage to gardens or decorations are nuisances which can give rise to complaints that effectively and expensively stop a contract if the potential problems have not been foreseen and eliminated as far as possible.

Good information for the occupiers can work wonders in gaining their co-operation, and where housing is being repaired it pays to hand-pick all the workmen because they are bound to get under the occupiers feet from time to time. On a well run contract the occupiers adopt the workmen and can create conditions that greatly increase the efficiency of working; on a badly run contract they will stop the work if they are sufficiently upset.

3.3 Specification

Contracts for repairs must be flexible about the extent of the work that is to be done, but they must be specific about the methods and standards to be achieved and the materials to be used. Other sections of this Guide give general information on what is needed.

3.4 Proprietary materials

Proprietary materials are often specified for repair work, because developments in repair materials are taking place all the time and some manufacturers are discouraging to people who enquire to discover the generic type of material that is offered.

This is not an ideal situation because it tempts specifiers to specify without knowledge. Specifiers must be knowledgeable about the types of material that are used in proprietary products and must ensure that the proprietary products they specify are suitable for the conditions they will be used in. They must demand to be advised about changes in specification or formulation made by the manufacturers and must accept manufacturers' test results only if they really understand what they mean. This does not imply that the manufacturer of the product is relieved of any responsibility for product performance, but, except in the case of suppliers who also apply their own products, workmanship is such a critical factor that it is difficult to ascertain whether the material is at fault if the repair fails to perform satisfactorily. A manufacturers' guarantee or warranty therefore will usually not be very helpful if a repair fails.

On large repair contracts, the consultants may have the resources to design and specify the repair materials in detail in consultation with the suppliers of specialist materials. They can then issue their specifications to proprietary suppliers as well as their contractors.

In this way the best use can be made of the consultants' knowledge of the requirements, combined with the suppliers' and formulators' knowledge of repair materials.

In any case, consultants must specify methods and materials rather than the performance to be achieved. Although contractors can be expected to give guarantees for perhaps 10 years for their work, the circumstances of repair work make it difficult to apportion blame if early failure occurs.

Both consultants and their clients have to understand that concrete repair is not yet an exact science that gives calculable results.

(See references 12, 16, 24 and 25 for further information on topics in this section).

4 Methods of repair

4.1 Repairing cracks

Inspection and testing of the structure will have established the type and cause of the cracking. The repair method that is suitable for each situation is discussed next.

4.1.1 Plastic shrinkage cracks

In situations where the concrete surface will be exposed to salt, these cracks can provide a reservoir of contamination close to the level of the reinforcement unless they are sealed before contamination can enter them. This only possible during or very soon after construction, and if these cracks are found at that stage they should be sealed with polymer-modified grout. (See Section 2.2.1.4.)

4.1.2 Plastic settlement cracks

Surface cracks in recently cast concrete can be sealed in the same way as plastic shrinkage cracks, or if another lift is soon going to be cast over them, they can be left untreated.

The water voids that form under reinforcement when concrete bleeds and plastic settlement occurs are difficult to find and very difficult to treat, but they are a potential cause of corrosion because the protection of the reinforcement depends critically on the cement coating around the bars being complete.

When bleed water collects under reinforcement, the whole of the underside over quite long lengths may be unprotected from rusting (and may also have very little bond with concrete).

By the time such situations are discovered it is usually too late to prevent rusting. (See Section 2.2.1.5)

Locating and filling water voids in completed structural members is very difficult and is almost impossible to do thoroughly. It is therefore extremely important to avoid this defect during construction by ensuring that the properties of the fresh concrete are correct for the way it is going to be used.

This bar has rusted only where it was not protected by cement paste.

4.1.3 Shear and transverse cracks

Cracks of this type which have been caused by stress *and which have not yet allowed significant contamination to enter or carbonation to take place,* can be sealed by injection with resin if they are inactive and do not move, or can be widened at the surface and sealed with joint sealant if they are active. (See Sections 2.2.1.2 and 2.2.1.3)

Whether these cracks need treatment at all depends on their width, the conditions of exposure, and whether the crack runs parallel with any adjacent reinforcement.

Where the cracks are not parallel to any adjacent reinforcement, they can be treated as shown on the following Table.

Crack width	Environment	Treatment if not contaminated
up to 0.50 mm	internal or salt free external	none
	maritime, salted highways, chemical plants etc	if dead, seal by resin injection if live, enlarge and seal as joints.
more than 0.50 mm	dry internal environments	none
	all other environments	if dead, seal by resin injection if live, which is more likely with cracks of this width, enlarge and seal as joints

Where reinforcement runs in two directions at right angles, cracks that are transverse to one bar will be parallel with another.

In such cases cracks should be sealed even if they are very fine (say less than 0.30 mm wide) because longitudinal cracks promote rapid rusting in most environments.

Normally, longitudinal cracks will be dead and can be sealed by resin injection. If measurements or the circumstances suggest that they are live, they must be widened and sealed as joints.

Resin injection is a well-developed technique undertaken by several specialist contractors. Low viscosity epoxy resins are used whenever a structural bond is needed between the faces of the crack. Alternatively polyester or acrylic resins can be used if it is necessary only to seal a crack against the ingress of moisture, but they have the disadvantages of lower strength and greater thermal contractions after hardening. The resin and hardener are usually premixed in a batch which is fed into the crack either under gravity or under applied pressure. The resin in fed progressively upwards starting from the lowest point so that air is expelled, though vacuum-assisted injection is also possible.

> Resin injection of cracks in concrete is a highly skilled process; its success depends largely on the experience of the operator. It is not a process that should be attempted by the general contractor.

4.1.4 Live cracks

Where a crack is live and continues to move with changing loads or temperatures it cannot successfully be glued together with resin. If the concrete is still uncontaminated the crack must be widened at the surface and sealed as a joint. (See Section 2.7)

Where the concrete is carbonated or contaminated with chlorides, it must be broken out and replaced, and a sealed movement joint formed at the same time. (See Section 4.2.4.)

The reason a wide sealing groove has to be made is obvious once it is realised that 25% strain – a reasonable amount to be accommodated by a flexible joint sealant – corresponds to a movement of only 0.125 mm in a 0.5 mm crack.

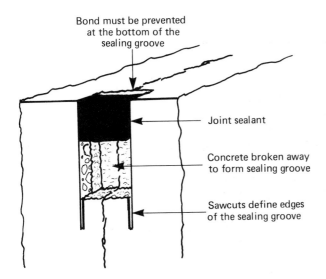

Bond must be prevented
at the bottom of the
sealing groove

Joint sealant

Concrete broken away
to form sealing groove

Sawcuts define edges
of the sealing groove

The sealing groove must follow the line of the crack and the sealant must be prevented from sticking to the bottom of the groove where movement of the crack could otherwise start a fatigue crack in the sealant.

The sealing groove can be made with a power hammer fitted with a sharp chisel or crack cutter, or with a fine high-pressure water lance operating at 5000 to 15 000 psi (300–1000 bar). It is difficult to follow a crack acceptably with a saw cut or hand-held angle grinder.

4.1.5 Longitudinal cracks caused by rusting

Cracks that run directly over reinforcing bars in such a position that they could not have been caused by shrinkage, plastic settlement, or thermal contraction, have been caused by the build up of rust forming on the reinforcement. These cracks are a symptom of deterioration which will eventually lead to spalling and complete loss of cover. They cannot be treated without removing and renewing the concrete cover. (See Section 2.2.1.1)

The soffit concrete in both these illustrations was removed by water blasting. (Soffit of a tunnel roadway).

4.2 Removing concrete cover

Where reinforcement is rusting enough to crack the concrete, the concrete cover must be cut away to the level of the affected reinforcement and usually up to 50 mm beyond it. Sometimes removing the cover is only too easy, and if the concrete sounds hollow when tapped the cover will probably fall away with very little trouble.

Normally, however, it is difficult to remove concrete cover, except where this is being done very locally, for example in the course of an inspection.

High pressure water blasting, using water at a pressure of 5000 to 15 000 psi (300–1000 bar), with or without an entrained abrasive is often the quickest way of removing large areas of high strength concrete and helps greatly in removing the bulk of the concrete behind the reinforcement.

Water blasting removes the weaker concrete preferentially, and leaves the remaining aggregate intact. If abrasive is entrained in the water jet, water-abrasive blasting is also able to remove most of the rust from the reinforcement that is exposed, though removing rust from the blind side is very uncertain and tends to depend on rebound from the concrete.

Power hammers are cheap to use and widely available. Even where water blasting has been used, power hammers will usually be needed for part of the work, but they have the disadvantage of shattering aggregate which is not removed. This damage can be minimised by ensuring that the tools are kept properly sharpened.

Preparation for concrete repairs to motorway support structure in Toronto.

4.2.1 Treatment of patch edges

If concrete is broken out from inside a large area of generally sound concrete, the edges of the area to be repaired have to be cut in such a way that the replacement material is properly contained by its surroundings and does not come to a feathered edge. This is necessary because the wedge which forms at a feathered edge can be prised off as a result of the compressive stresses which are unavoidable when the temperature changes. The thermal and elastic properties of the patch can never exactly match those of the remaining original concrete although materials should be chosen to get as near a match as is possible. (See Section 4.6.1).

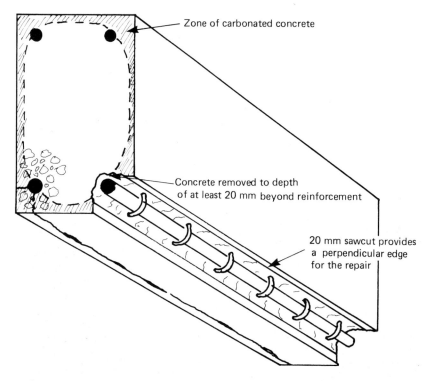

Zone of carbonated concrete

Concrete removed to depth of at least 20 mm beyond reinforcement

20 mm sawcut provides a perpendicular edge for the repair

Some repair instructions recommend that the concrete edge should be under-cut, but this brings the converse problem that the *original* concrete now has feathered edge and may be prised off. Under-cutting is also quite difficult to achieve in practice because small pieces of the overhang keep breaking away next to the concrete which is being removed.

By far the best and most practical way is to make a *perpendicular* sawcut all round the area to be removed to a depth of at leasty 15 mm beforehand. The face of the cut must be roughened slightly (for example with a needle gun) to help the replacement material to bond with it.

This way there are no prising forces and the repair looks much tidier too.

The only satisfactory alternative to a sawcut for defining the edges of a patch is the very fine high-pressure water jet developed specifically for 'precision' cutting rather than removing concrete.

Removing concrete cover over local areas of damage can lead to very unsightly results when the concrete has been patched and it is sometimes worth extending the repaired area to some line where the boundary of the new work will fit in with a feature of the structure. Repaired patches in any position will tend to look much better if the saw cuts that define the patch are accurately vertical and horizontal or on a radius or circumference if the structure is curved. The eye instantly picks out anything that looks crooked or accidental in a man-made structure.

Where concrete cover has to be removed very near an external corner it is best to extend the removal a short distance along the flanking face and use formwork to re-form the feature when the patching is done. It is difficult to avoid damage to external corners when removing concrete close to them in any case.

4.2.2 Depth of removal

Where reinforcement corrosion is the cause of the problem, obviously concrete must be removed to a depth that includes all the affected reinforcement and leaves some room for replacement behind it as well. (Remember there may be more than one layer of affected reinforcement). How much further it is necessary to go is less obvious.

Some sources recommend that *all* carbonated concrete should be removed, regardless of how far it extends beyond the reinforcement. This is not necessary if the reinforcement is not in contact with carbonated concrete at any other point in the member. Provided the reinforcement is surrounded by new uncarbonated concrete which extends to give an adequate thickness of cover (see Section 2.4) in every direction that carbon dioxide can come from, there is no point in going further. Carbonation does not spread from carbonated into uncarbonated parts unless new carbon dioxide reaches them.

Where chloride contamination is the cause of the problem, there are more facts to consider. If enough chloride contamination to cause corrosion comes from the original mix (see Section 2.5), no limited amount of further removal will give *complete* safety from further attack, because unlike carbonation, chloride can spread from contaminated concrete into the new concrete. How quickly it spreads depends on many factors but if the new concrete is of sufficiently high quality (i.e. high cement content, low water/cement ratio and of good compaction) it is likely to provide protection to the reinforcement for a very long time if the cover thickness *all round* is adequate. Chloride penetration to the reinforcement from the original concrete is likely to be no faster and probably a great deal slower than from an external salty environment, unless the chloride is carried through the concrete, by water flow for example.

The extent (rather than depth) of removal is discussed in the next section.

Total failure of soffit at a filling station canopy in Sharjah.

Chloride contamination which has come from a salty environment decreases as the distance from the surface increases and at some depth may be insignificant (say less than 0.1% of cement in uncarbonated concrete). In this case it would often be possible to remove the contaminated concrete altogether. If the cover is then replaced with concrete which is highly impermeable to chloride, the cause of the original deterioration will have been completely removed. (See Section 2.5)

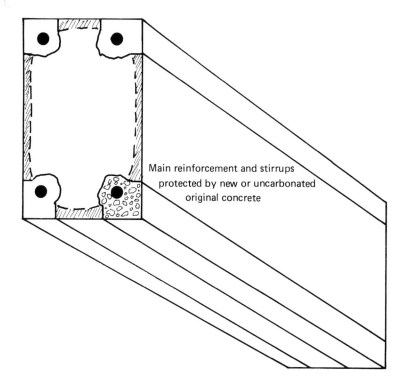

Main reinforcement and stirrups protected by new or uncarbonated original concrete

4.2.3 Lateral extent of removal

Cracking or spalling from reinforcement rusting often occurs quite locally on a structure. It may start where the cover is especially thin, where the concrete has cracked and allowed chloride carbon dioxide to penetrate to the reinforcement, or at a site of local contamination such as salt run-off from a bridge deck.

If the damage has a purely local *cause*, only the damaged area of the concrete plus the length needed to bend any new reinforcement needs to be removed, (see Section 4.4) but if the damage is local only because it happens to be for the moment the worst spot in an area which is generally in danger of deterioration, repairing locally may move the trouble to some other position.

This can happen when a strongly corroding trouble spot in a larger area of non-passive reinforcement acts as a sacrificial anode and gives cathodic protection to the remainder of the reinforcement which would otherwise also be rusting.

This condition is most likely to be found in areas of concrete which are salt-contaminated and at the same time damp enough to have comparatively low electrical resistivity. Unwaterproofed highway bridge decks in regions where de-icing salts are used are a common example that is very well documented. (See Section 1.2).

In such cases, it will often be necessary to remove contaminated concrete in areas where the reinforcement is not yet rusting to prevent it from starting to rust when the currently rusty areas are repaired. Electrode-potential surveys (see Section 2.12) and a thorough examination of the condition of the structure should show whether this is likely to happen. This type of corrosion mechanism is explained in more detail in Sections 1, 2 and 6.

4.2.4 Removing concrete at joints

Joints in structures are especially vulnerable to deterioration for several reasons:-

1. They can be difficult to construct and the concrete at a joint may lack compaction.
2. They may act as paths for the entry of salty water or carbon dioxide.
3. They may fail to work as joints forcing the concrete may crack at an adjacent plane of weakness (e.g. at the end of the dowel bars or the fin of a water-stop).
4. They may be intended not to be active joints (construction joints for example) but may subsequently become active without having any provision for sealing.

Construction joints which open and behave as unintended movement joints can be widened and sealed at the external surface if this is done as soon as they are noticed and before contamination can enter. If they are already contaminated or if the

Deterioration often starts at construction joints.

reinforcement is already rusting, they must be repaired by removing the affected concrete and reconstructing the joint. The reinforcement will be needed again. Take care not to damage it (further) when the concrete is removed!

Removing the concrete 'cover' at a joint is often a matter of removing concrete both on the external surface and the internal joint face, and it usually results in all the concrete in the immediate vicinity of the joint having to be removed. Water blasting is again usually a good way of doing it.

If it is necessary to transfer shear forces at a joint, the load will have to be carried by dowel bars which are accurately perpendicular to the joint plane and free to slide within the concrete on one side of the joint at least. Dowelled joints which are aligned with the accuracy that is needed for them to slide freely without stressing the concrete must be pre-assembled to close tolerances and installed with great accuracy in the formwork before the concrete is cast.

In new construction dowelled joints must be accurately aligned. Remember that misaligned joints are a very common cause of failure in concrete structures.

When a joint with shear-load transfer is repaired, it is often necessary to remove the joint assembly completely. This means that all the concrete may have to be removed as far as 0.5 m on either side of the joint, though the reinforcement must of course be left in place so that the recast concrete can be tied-in to the existing concrete.

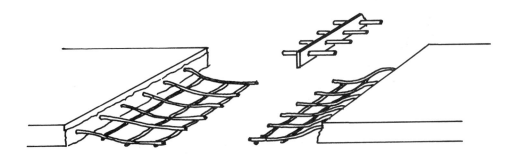

Attempts to repair dowelled joints by removing the concrete on one side only and straightening the projecting ends of the dowel bars before recasting seldom succeed. It is virtually impossible to achieve the required accuracy of alignment in this way.

Dowel bars can be installed successful in slots cut from the concrete surface. Carefully sawn, they can ensure excellent alignment in both line and level. (See Section 4.4)

4.2.5 Safety from collapse

Removing concrete cover can cause a member to collapse. Structural repairs must never be undertaken without the supervision of a competent structural engineer.

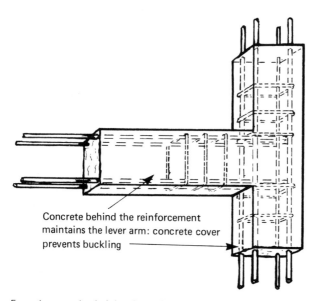

Concrete behind the reinforcement
maintains the lever arm: concrete cover
prevents buckling

Even the most basic joint depends on the concrete cover for its strength.

> BEFORE REMOVING ANY CONCRETE FROM A LOAD-BEARING STRUCTURE CONSIDER CAREFULLY WHETHER THE CONCRETE YOU PROPOSE TO REMOVE IS PROVIDING ESSENTIAL SUPPORT FOR THE STRUCTURE.
>
> IF IT COULD BE, SUPPORT THE STRUCTURE FIRST!

4.3 Cleaning reinforcement

When the concrete cover has been removed the reinforcement can be inspected. If water-blasting was used, it is likely that most of the rust will have been removed with the concrete but the roughness of the surface will still show where the original rust was, and some rust will usually remain on the blind side of the bars. In most cases much of the rust will remain.

Reinforcement is often slightly rusty before the concrete is poured. This rust (if salt-free) is harmless and looks quite different from rust which has formed subsequently. The pre-construction rust will be firm and lightly coated in cement: the post-construction rust will have stained the concrete and will not have the normal cement film on the surface. Flash rusting following water blasting is harmless if the chlorides have been removed completely.

If the cause of the damage which is being treated is local chloride contamination it is essential to remove *all* rust from the reinforcement. The SA $2\frac{1}{2}$ grade of the Swedish painting standard is a suitable level to achieve. This is the only way to ensure that the steel-concrete interface is free from chloride contamination.

Water-abrasive blasting should be used to remove the rust. The combination of abrasive to remove solid contamination and water to dissolve chlorides is one of the best ways of ensuring a chloride-free surface for the reinforcement and probably the only way of removing chlorides from the pits in rusting reinforcement. Fortunately pits nearly always face the outside of the unit, but enough concrete must be cut away on the blind side of the reinforcement to allow room for water-abrasive blasting. The space needed will probably be no more than is needed anyway for concrete cover to resist the migration of chlorides towards the steel in the repaired area (see Section 2.5)

If the cause of the damage is carbonation, rust removal is less critical and it will be sufficient to remove any loose rust which might prevent the reinforcement from being coated with a layer of firmly adhering cement paste (or other material if the repair is not to be a cementitious one).

4.4 Adding reinforcement

If rusting has reduced the cross-sectional area of reinforcement by more than 20%, extra reinforcement must be added before the repair is made good.

The usual method of adding reinforcement is to lap the weakened bars with additional bars to restore the cross-sectional area to its original value. The lap lengths over the undamaged parts of the bar should be the same as those required for new construction by the appropriate code.

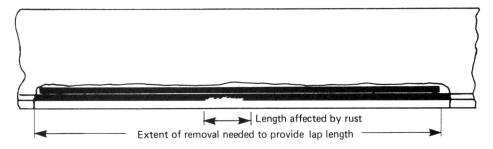

Length affected by rust

Extent of removal needed to provide lap length

It is usually necessary to remove far more than the damaged concrete to expose enough of the reinforcement to make satisfactory lap joints.

In some situations it may be better to add reinforcement by drilling into the concrete and bonding new bars into the drilled holes with epoxy resin. Examples are extra reinforcement near joints, dowelling to replace rusted links in columns, or reinforcement to anchor an extra thickness of cover. Using this method avoids electrical continuity between new and old bars and is therefore preferable when adding extra cover. This type of anchorage is best if it is drilled at an angle to the direction of the applied stress so that the reinforcement is locked into the hole as well as bonded.

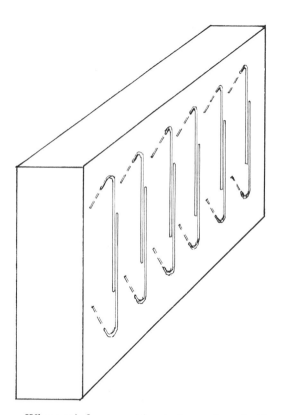

Where reinforcement has to be anchored at both ends it is usually more convenient to install it as two separate pieces which are lapped or welded together after each has been anchored. Installing both ends of a single reinforcing bar into drilled holes is sometimes impossible and usually awkward enough to invite damage from an attempt to force it into place.

If the surface of the concrete can be cut conveniently, reinforcement can be anchored in slots cut from the concrete surface. Dovetail slots cut at an angle to the stress helps to anchor the bars.

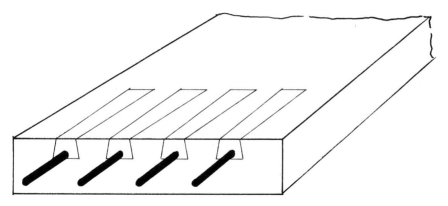

Dowel bars and ties can be anchored in dovetail slots cut from the surface.

A similar method is sometimes used for installing new load-transfer joints in concrete paving.

4.5 Coating reinforcement

There is controversy over the need to coat reinforcement in areas that are being repaired.

The coatings that could be used divide broadly into:

1. cement slurry
2. cement slurry modified with polymer or latex emulsion
3. epoxy resin (with or without alkaline admixture)
4. inhibitive primer (for example zinc chromate primer)
5. zinc-rich sacrificial primer.

Alternatively, the cement paste from a well-designed cementitious repair mix may protect the reinforcement better than any separately-applied coating.

Whether a coating should be used, and if so, what type, depends partly on the way the concrete is to be replaced and this is discussed in the next sections.

When the repair is cement-based, it is now believed the barrier type of coating may do more harm than good by shielding the reinforcement from the alkaline cement paste without giving the protection of a *perfect* barrier in its place.

Modified cement slurries may dry too quickly to be effective in repairs where forms have to be fixed after the coating is done, but they are suitable where the delay is short.

A warning has to be given to anyone who recommends a coating as an extra precaution in cases where it is very difficult to clean the reinforcement effectively. A coating is not an alternative to removing chloride contamination from the reinforcement nor will it prevent corrosion from being caused by chlorides which are already present on the reinforcement.

Sacrificial zinc-rich primers have recently found favour because it is thought that they can give the reinforcement some initial protection, though this has not yet been proved by long-term experience.

Reinforcement coated with fusion-bonded epoxy resin is sometimes used to repair concrete in very aggressive environments. If the coating remains undamaged it should effectively protect the reinforcement from corrosion. Where there is a risk of damage to the coating, it is important that no uncoated reinforcement should be used near it, because there is then the danger that a large corrosion current generated by an uncoated steel cathode could produce intense local corrosion at anodic breaks in the coating (See Section 1.2).

Stainless steel or non-ferrous metals should not be used in contact with steel reinforcement because they can increase the corrosion rate at remote anodic areas by galvanic action.

Non-metallic reinforcement is a comparatively new development. Its durability and performance are still uncertain, but it cannot cause corrosion by contributing to galvanic reactions.

Epoxy painted reinforcing bars.

4.6 Replacing concrete

Areas of concrete which have been cut away can be made good in five basically different ways:

1. by recasting with concrete;
2. by patching with trowel-applied cementitious mortar;
3. by spraying-on new concrete;
4. by prepacking dry aggregates which are subsequently grouted;
5. by patching with trowel-applied resin-based mortar.

Some of these methods have special advantages in certain circumstances.

4.6.1 Material properties and performance

Replacing only part of a structural member with new material will affect its behaviour if the repair material has different properties from the original concrete.
It is necessary to understand what the effect is likely to be.
The most important properties to consider are elastic modulus, coefficient of thermal expansion, impermeability and strength.

Table of material properties

Material	Compressive strength N/mm^2	Elastic modulus kN/mm^2	Coefficient of Thermal Expansion $\times 10^{-6}/C$
Concrete with limestone aggregate	15 to 70	20 to 40	7 to 9
Concrete with siliceous aggregate	15 to 70	20 to 40	12 to 14
Sand–cement mortar (about 3:1)	10 to 20	—	10
Sand-filled epoxy resin mortar	50 to 100	0.5 to 20	20 to 50
Sand-filled acrylic resin mortar	—	10 to 25	5 to 15
Polymer-modified cement mortar	10 to 60	1 to 30	8 to 15

Different concrete mixes have slightly different properties from each other, but resin based materials have very different properties from any kind of cementitious concrete. Generally speaking resin-based materials have an elastic modulus of about one tenth of that of concrete and a coefficient of thermal expansion of more than five times that of concrete. The strength of resin-based materials in compression is usually higher than the strength of concrete, and in tension it is much higher.

The effect of these differences depends on where the materials are used. Resin mortar used to make up part of a compression member will not carry its share of the load until it has deflected much more than the concrete. This may cause eccentric loadings . . .

Concrete —————▶   ◀————— Resin mortar

Concrete

Resin mortar

or it may produce tensile stresses when the lower modulus material spreads under an applied load.

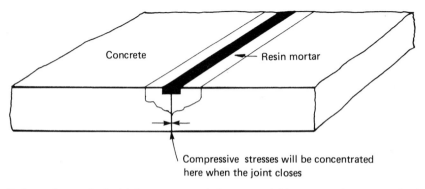

Concrete   ◀— Resin mortar

Compressive stresses will be concentrated
here when the joint closes

Resin mortar repairs to joints may concentrate unacceptable compressive stresses in the adjacent concrete if insufficient space has been provided for expansion movement.

For repairs which need close-fitting packing, for example under bearings, only resin-based materials allow the packing to be poured into place.

Recasting repaired areas with concrete avoids many of the problems caused by adding material with different mechanical properties, though it does raise problems of its own, one of which is that concrete shrinks significantly when it hardens.

Whenever possible, cementitious materials should be used for repairing concrete structures. If it is not possible to use concrete because there is insufficient room or because it is impractical to provide formwork, use cementitious mortars or, if the area is a large one, consider sprayed concrete.

Cementitious mortars and even concrete may be modified by the addition of polymers, but cementitious materials of some sort are always preferable to resin mortars for concrete repairs if the circumstances make it practical to use them.

4.6.2 Recasting with concrete

4.6.2.1 Preparing the old concrete surface.

When existing concrete is repaired by adding new concrete, the old concrete must be very clean so that the new cement paste can bond with it. The old concrete must be saturated so that water needed for curing is not sucked from the new concrete but it must also be dry on the surface so that the water/cement ratio at the interface is not increased.

The existing concrete must be saturated in depth, not just at the surface, because the excess water needed for curing the new concrete must be available for *at least* the first 24 hours after it has been cast. Old concrete which is saturated only near the surface will start to suck water from the new concrete within a few hours of it being cast and this will prevent full hydration of the cement at the critical interface between the old and new concrete. The simplest way of ensuring that the old concrete is in a suitable condition to bond with the new concrete is to saturate it with water sprays for at least 24 hours before the new concrete is cast, and then to allow the *surface* to dry off naturally for an hour or two before the new concrete is cast. Obviously the time this takes will depend on the weather.

The time and cost of preparing the old concrete in this way is negligible by comparison with the time and cost of the other operations which will have been needed to bring the repair work to this point, but proper preparation is of critical importance.

4.6.2.2 Bonding coats

There is controversy about the need to use a bonding coat between the old and new concrete, and tests have produced conflicting evidence. This is probably because bonding is very sensitive to the precise conditions at the interface and these are difficult to standardise even under test conditions.

Some factors are reasonably well understood. A film of cement paste is needed between the old and new work to coat all the aggregates and the hardened cement paste of the old concrete surface. This film can be separately applied cement grout, or it can be grout from the new concrete. Its intimate contact with the old surface is important: grout must be well worked into the concrete surface, and concrete must be very thoroughly vibrated.

> Notice how well unwanted mortar bonds with the concrete driveway or garage floor where it was mixed when the garden wall was built – not a result of nature's cussedness, but of the intimate contact that is ensured by vigorously scraping cement mortar over the existing concrete surface while it is being mixed.

Some authorities recommend that bonding grout containing a polymer additive should be used to coat the old concrete, but there is no proof that this has any special advantage. Polymer-modified bonding grouts have a short drying time (normally less than 30 minutes) and cannot be used if there is much form-fixing to be done before the concrete can be cast. Cement grout allows rather more working time, but if the form fixing is lengthy and complicated (or cannot be set-up in advance for rapid re-fixing) the use of a bonding coat will be impossible, because it will be dry before the concrete can be cast.

Where bonding coats are used they should be mixed in high-shear-action mixers to avoid excessive air entrainment.

Water-compatible epoxy resins were at one time widely used as bonding coats. Although cementitious bonding coats are now generally preferred, epoxy bonding coats have two special advantages. Firstly they can be formulated to have long open times, which makes them suitable for use in hot climates, or when formwork has to be fixed after the bonding coat has been applied. Secondly, they may provide a more effective barrier than cement grouts against the migration of chlorides from within the concrete.

Although they are water compatible, epoxy bonding coats are applied to a dry concrete surface.

4.6.2.3 Concrete mix design

The concrete mix used for recasting must be capable of producing highly imperme-able concrete and must of course be free from contamination.

Mix design for concrete repairs needs skill and experience. In general it should follow normal good practice but there are some special requirements which do not apply to the same extent to mix design for normal concrete.

1. Ideally the repair mix should be made with the same type of aggregate as the original concrete to minimise thermal stress, but if the original aggregate was of poor quality, good material of the same geological type is preferable. Because a rich mix will be used, special care should be taken to ensure that aggregate will not be in danger of reacting with the alkali from the cement.

It is usually necessary to use a smaller maximum aggregate size for repairs because the space tends to be restricted.

2. The repair concrete should be very low in free water to minimise stresses caused by drying shrinkage. The water:cement ratio should not exceed 0.45. In some situations it may be helpful to add shrinkage-compensating admixture to the mix. These admixtures work by causing slight expansion to offset shrinkage and thermal contraction.

3. The repair concrete should have a high cement-paste content to ensure that both the interface and the reinforcement can be properly coated, and the sand grading must be chosen to keep bleeding to the absolute minimum, especially for soffit repairs where bleeding can lead to complete separation between old and new concrete.

The conflicting requirements of low shrinkage and high paste content can be reconciled to some extent by the use of a superplasticiser to give a mix of medium workability at a very low total water content.

Additives such as ground granulated blast furnace slag, pulverised fuel ash or microsilica can be used in recasting mixes to increase impermeability in the same way as in new concrete.

4.6.2.4 Compacting recast concrete

Where concrete is replaced by recasting, it must be made to flow into place to exclude all air voids. This is no more critical than the need for compaction in new work, but it can be much more difficult to achieve because the space for placing the concrete and using internal vibrators is often very restricted. The need to keep shrinkage to a minimum often makes it advisable to use superplasticisers to produce even medium workability concrete and often rules out the use of very workable self-compacting mixes. The most practical way of achieving good compaction is by placing the concrete in small amounts and vibrating it conscientiously as the work proceeds. External vibrators on the formwork can be very effective for inducing flow and compacting concrete in awkward places.

Where soffits are recast it is sometimes possible to drill through the slab and place the concrete from above. If cores are cut as part of the examination of slabs with defective soffits, thought should be given to taking cores from positions which can be used later for placing new concrete.

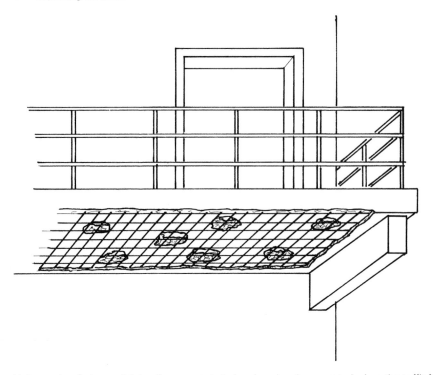

Holes cut in a balcony slab to allow concrete to be placed and compacted when the soffit forms have been fixed.

Alternatively, soffit formwork can be extended to provide a series of hoppers at the open sides for feeding superplasticised 'flowing' concrete under gravity head, or the concrete can be pumped to the furthest point of the void through a pipe which is gradually withdrawn as concreting proceeds. If this method is used, the formwork must be designed to withstand the pumping head. Forethought about profiling the surface when removing concrete from damaged areas can help greatly in providing ways of getting new concrete into position and the air out (see Section 4.2).

Bleed water can cause complete separation between the repair and the original concrete when soffit concrete is replaced. Great care is needed to design mixes which are as free as possible from bleeding.

Whichever method is used, any void which is confined on all sides must be filled from the lowest or farthest point to ensure that air is driven out as the concrete advances, and it may be necessary to fit the formwork with removable baffles to force the concrete to move in the required direction.

This soffit repair in the Middle East failed again after a very short while.

4.6.2.5 Formwork for repairs

Formwork for recasting repairs must be very rigid to prevent the new concrete from sagging away from the old concrete under its own weight, to withstand pumping forces if concrete or grout are placed in this way, and to take the forces of clamped-on external vibrators. Heavy gauge steel forms are ideal because as well as being rigid, they allow heat to escape from the hydrating concrete and thus reduce the contraction stresses when the newly hardened concrete cools, though craneage will be needed to handle them in anything other than very small sizes.

If there is enough repetitive work to justify the cost, permanent formwork of glass-reinforced plastics can be useful in some applications, for example repair of piles.

Forms must be well supported as well as being rigid. Soffit forms can sometimes be supported with props from the ground or a lower floor, on tall structures they can be supported with falsework fixed to the structure itself.

The support provided to serve as access for inspection and preparation of the old concrete may be suitable for supporting formwork and this should always be considered when the access system is designed, but the need for more rigidity and sometimes more strength than is normally needed for access must not be overlooked.

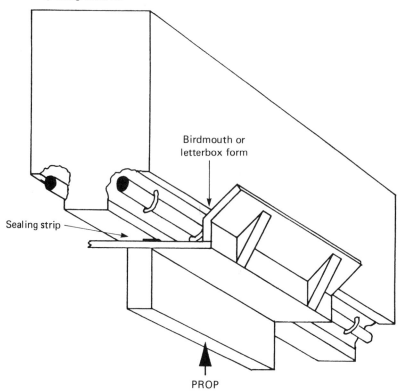

Windows or hoppers usually have to be incorporated in forms for soffits or vertical faces so that concrete can be fed into the void, but the need for extra openings to allow poker vibrators to be used is sometimes less obvious. Steel is often the most suitable material for making rigid forms that have to be fitted with a number of optional alternative openings which can be closed as the pour proceeds.

4.6.3 Patching with cementitious mortar

Small areas of concrete up to about 50 mm thick are normally replaced with trowel-applied mortar rather than concrete that contains separate aggregate. Cement-based mortars usually use relatively coarse sand at a ratio of about 3:1 by weight of cement.

Because of its high specific surface, mortar of a consistency suitable for trowelling usually does not contain enough cement paste to coat the old concrete surface or the reinforcement adequately and a bonding coat of cement grout must be applied to the old concrete surface and to the reinforcement unless it has not been treated as recommended in Section 3. With trowel-applied mortar, using a bonding coat creates no special problem because there is no formwork to fix and the mortar can be applied immediately after the bonding coat.

Polymer admixtures, usually based on styrene/acrylic or styrene/butadiene emulsions can be used in both the mortar and the bonding coat and may improve the performance of the repair. These admixtures are usually supplied as water-based emulsions but some are available as dry powders and this allows them to be incorporated in pre-packed repair materials which overcome site batching problems. About 10% (dry weight) of polymer by weight of cement is the smallest amount that gives a significant increase to the adhesion and tensile strength of the mortar. Specially formulated cementitious patching mortars are available commercially and combine a range of admixtures to give the mortar particular properties like good adhesion, impermeability and increased flexibility.

If the mortar is mixed for too long, or in an unsuitable mixer, polymer admixtures can cause excessive air entrainment. Short mixing in a high-shear mixer is best. The old concrete should be saturated but clean and dry on the surface for the reasons explained in Section 4.6.2.1. Even under these conditions, bonding coats will dry quickly if they contain polymer admixtures and in hot dry weather the work must be organised so that the application of mortar follows the bonding coat within a few minutes. The bonding coat will prevent bonding if it is allowed to become dry.

The repair mortar must be firmly punned into place so that as much air as possible is excluded, and the surface must be closed by trowelling. It is not possible to achieve the same densities with sand/cement mortars as with recast concrete and the protection of the steel will rely largely on the effectiveness of the bonding coat that was applied to it.

Specially formulated repair mortars have much better properties than simple sand/cement mortars and can give protection equal to or better than recast concrete.

4.6.4 Sprayed concrete

Replacing concrete on damaged structures by spraying on new concrete is a simple technique which is widely used in high-volume concrete repair work. The impermeability of sprayed concrete can be increased by the use of latex additives or silica fume as an admixture.

4.6.4.1 Dry process spraying

This is the spray process most often used for repair work. Cement, additives, and aggregates up to 20 mm in size are batched and dry-mixed in conventional plant and then sprayed by air-propulsion through the nozzle where water is added by the nozzleman to give the right consistency for the concrete to stick to the sprayed surface without slumping or rebounding excessively.

The aggregate:cement ratio used is usually 4:1 or a little richer. The strength achieved is typical of good quality structural concrete, though it varies more than the strength of premixed concrete because the water content is constantly being modified by the nozzleman to give the best consistency for adhesion to the surface.

4.6.4.2 Wet-process spraying

This spray process uses normal pumped concrete which is propelled through a nozzle by a high pressure air stream. The mix characteristics are unchanged by the spraying process and the strength is therefore more consistent than with the dry process.

Any mix which is pumpable can be placed by the wet process, but rate of delivery has to be high enough to match the throughput of the pump and that makes the process unsuitable for small areas or complicated shapes.

Sprayed concrete must be reinforced with small-mesh small diameter reinforcement fixed near the concrete surface or with random fibre reinforcement sprayed as part of the concrete mix to prevent the development of cracks when the concrete shrinks.

Spraying is most suitable when large areas of relatively thin concrete (30 to 60 mm) have to be applied, for example when it is necessary to restore or increase the cover over reinforcement which is at or near the carbonation depth of the concrete. If existing cover is being increased, the surface of the old concrete must be cleaned and roughened first.

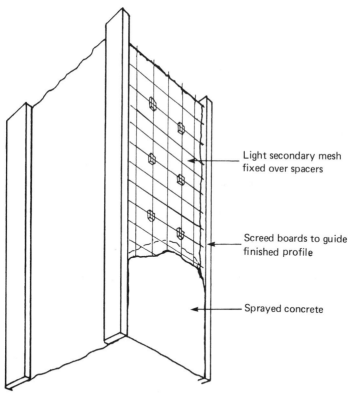

Light secondary mesh
fixed over spacers

Screed boards to guide
finished profile

Sprayed concrete

4.6.5 Prepacked aggregate concrete

Shrinkage can be minimised in re-cast repairs by suitable mix design, but it cannot be eliminated altogether. Repairing with prepacked and grouted aggregate can produce a system which is virtually free from bulk shrinkage because the prepacked aggregate fills the whole of the repair space and what shrinkage there is takes place without moving the aggregate.

The old concrete should be prepared in exactly the same way as for recasting concrete, but the formwork can be simplified a little because there is no need for openings to allow poker vibrators to be used.

In prepacked construction single-sized coarse aggregate is packed behind the forms so that it fills the void completely and grout is pumped from the farthest or lowest point of the void to fill the spaces between the aggregate.

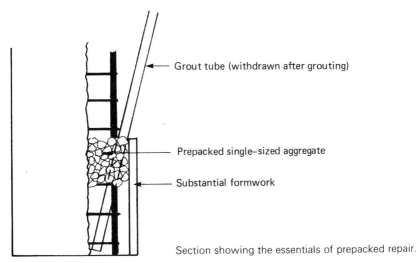

Grout tube (withdrawn after grouting)

Prepacked single–sized aggregate

Substantial formwork

Section showing the essentials of prepacked repair.

Where the void is large enough to allow it, 20 mm aggregate and sand/cement grout should be used. For smaller voids 10 mm aggregate can be grouted with neat cement grout or cement/pfa grout. Bleeding and blocking can be problems in grouted construction and success is very much a matter of trial, error, experience and the use of the correct equipment for mixing and pumping the grout. Very few contractors have any experience of prepacked aggregate construction and it is a job for specialist contractors unless it is being used on a very small scale.

Trials are almost always necessary to identify and solve practical problems which arise with this type of construction.

4.6.6 Curing

All types of cementitious repair, whether or not they contain polymer admixtures, need thorough and continuous curing for at least 3 days in temperate weather or 7 days under hot drying conditions. This is essential both to develop the impermeability of the repair materials and to reduce the stress of drying shrinkage to a minimum while bond strength is developing. Drying from the external surface is especially harmful to the bond because it produces a tendency for the edges of the repair to curl away from the old concrete as a result of differential shrinkage, like a sandwich that has been left out in the sun.

Initial curing can be provided by leaving the formwork in place. This must be followed by covering the repaired area with wet fabric which is itself covered by polythene sheeting to keep the water in. Sprayed-on curing membranes, which can keep most of the original water in place but cannot provide any reserve of extra water to make up for any that escapes, are not likely to provide enough curing during the critical early stages, though they can help to slow down drying after the wet curing has been removed.

Alternatively, curing can be provided by keeping the repaired concrete wet with water sprays if chloride-free water is readily available and good supervision ensures that spraying is continuous.

4.6.7 Patching with resin mortar

For very small repairs which are less than about 15 mm thick, mortar made from epoxy or acrylic resin mixed with fine sand can be applied by hand with a trowel. The old concrete should be as dry as possible, and the surface should be clean and completely dry.

Priming the surface with the liquid resin hardener mixture (without sand) helps to give good bond, but the mortar must then be applied before the resin hardens. The hardening time varies with the resin formulation, but can be as long as several hours if necessary.

Patching which is thin enough to be appropriate for repair with resin mortar is not likely to extend down to the reinforcement, but if it does, the reinforcement must be free from chlorides and rust, for example water-abrasive blasted (see Section 4.3) to Swedish SA2½ or, better, completely protected with liquid resin before the mortar is applied because there will be no alkaline environment to passivate it. Resin mortar patches do not need any form of curing and this can be a justification for using resins in some circumstances.

All resins suitable for concrete patching need either a reactive hardener (epoxy or acrylic resin) or a catalyst (polyester resins) to make them harden. Reactive hardeners must be proportioned within 10% of the correct amount to avoid a considerable reduction in material properties. For this reason they should be mixed in whole batches which are obtained pre-proportioned from the supplier. There is more tolerance with catalytic hardeners where changing the proportion of hardener affects the speed of hardening rather than the material properties.

Epoxy, polyester and acrylic resins can all be supplied as sand-filled mortars which harden through chemical action, but epoxy resins are by far the most common and are the only resins that can be used to make mortars with structural properties suitable for load-bearing applications.

There are many different kinds of epoxy resin. Some are solids at ordinary temperatures, some are high viscosity liquids, and those used in concrete repair are relatively low-viscosity liquids. Any of these resins has to be mixed with the correct quantity of a suitable hardener to convert it into a solid, and, once solid, epoxy resins cannot be melted-down and reformed, though they will burn, and temperatures as low as 50°C reduce their properties.

As well as many different kinds of epoxy resin, there are even more different materials that can be used as hardeners, and each has a marked influence on the final properties of the hardened resin.

Occasionally, resins are thinned with solvents or diluents; solvent free resins are often supplied ready-mixed with fillers which can range from fine powders or sand to fine graded aggregates of up to six times the volume of the resin.

The range of materials that can be made for different purposes means that the design of resin mortars (usually called 'formulation') is a skilled business which is usually done by specialists who supply ready-formulated products for specific types of application.

Skill and experience are also needed for the successful application of epoxy resin materials. The materials have to be proportioned accurately, preferably in whole batches which have been supplied pre-weighed, and mixed thoroughly in a high-shear mixer. They have to be applied within a very limited time before they harden, they have to be handled cleanly to avoid contamination of both the resin mixtures and the people working with them, and they can be removed only before they have set and then only with special solvents.

Epoxy resins can cause skin irritation and the hardeners used with them can be powerful irritants. These hazards can be avoided by taking the precautions recommended by the manufacturers. Advice on safe handling is generally enclosed with every consignment of material.

High temperatures shorten the working life of the liquid resin-hardener mixture and the setting-process itself gives out heat, so that a large batch of resin-hardener mixture has a shorter working life (pot life) than a small one. At low temperatures it can be difficult to get the mixture to harden quickly enough to remain in place where there is a tendency for it to slump away from its intended location, especially if the volume used is small. Resin mortars do not need to be cured but after initial hardening their full strength may take some hours or even days to develop.

Resin mixtures can be made fluid enough to flow into place by gravity so inaccessible places can be filled with them and compaction is unnecessary. This is a great advantage of resin mixtures over cement mortar or concrete and makes them valuable for packing bearings, bedding rails and so on, but any shims or packing that were used to locate the parts must be removed first because the resin is relatively compressible (compared say with steel) and will not take over the load from packing which is left in place.

Epoxy resins have several features which make it advisable to use specialists to apply them as well as supply them, but their special properties make them virtually irreplaceable for some repair applications.

(See references 8, 12, 13, 16, 17, 18, 22, 24 to 31 for further information on topics in this Section).

5 Surface treatments

Surface coatings like paints and varnishes, or penetrating treatments which are absorbed in the outer skin of concrete, can slow down the rate of deterioration in three basically different ways:

1. by increasing the resistance of the concrete to chloride penetration
2. by increasing the resistance of the concrete to carbon dioxide penetration
3. by reducing the moisture content of the concrete.

Coatings are also about the only reliable way of hiding the patchy appearance of concrete which has been repaired in different places. If a coating is needed for reasons of appearance it should be chosen to have protective properties as well.

The life of the coatings themselves is difficult to predict, but certainly they will not last nearly as long as a durable concrete surface would and it must be accepted that if concrete is coated for any reason, recoating it from time to time will be an unavoidable maintenance expense for the owner. Nevertheless, coating concrete may be an option which could give better value for money than the alternative of carrying out a repair in some circumstances, and painting may be necessary to give extra protection to repair in others.

In general, pigmented coatings give much better protection and are more durable than unpigmented coatings.

See the table of waterproof treatments at the end this section.

5.1 Coatings to resist salt and water

Salt penetrates into concrete from the surface only when it is dissolved in water: even in the finest form, it cannot penetrate as a solid. So the basis of a coating to resist salt penetration is one that will resist the penetration of water.

Any coating that will stick well to the surface and form a waterproof skin is able to reduce the penetration of salty water into the concrete and in severe exposure it may be wise to give this kind of protection to new concrete to increase its durability.

When concrete is already showing signs of deterioration and tests show that enough salt is present at the reinforcement to allow it to rust, adding a waterproof coating will not help by reducing the possibility of further salt penetration: it is too late for that to make any significant difference.

Waterproof coatings allow water-vapour pressure to build up behind them especially if water can get into the concrete from another face. This build up of vapour-pressure can cause the coating to blister and peel off unless adhesion to the concrete is very good. A similar problem can arise when resin mortars are used for trowel-applied repairs on concrete. This problem can be overcome (or reduced anyway) by using a priming coating which penetrates into the surface of the concrete and gives a foothold for the surface coating.

A common but special use of coatings is the waterproofing of concrete bridge decks before they are surfaced with a wearing course. These coatings must be able to withstand both the laying temperature of bituminous materials and puncturing from the aggregate when the surfacing is rolled.

5.2 Coatings to resist carbonation

Carbon dioxide (and other acid gases in industrial environments) penetrate into the concrete mainly as gases. Because of this relatively dry concrete carbonates more quickly than concrete which is damp or wet, because the pores contain little water to obstruct the entry of gas but enough to allow the gas to dissolve.

Coatings to resist gas penetration do not need to waterproof the concrete, but they usually do to some extent because they tend to block the pores. However coatings *designed* to resist water penetration by being water-repellent work by altering the surface tension in the pores rather than by blocking them.

A surface coating with low gas permeability can resist carbon-dioxide penetration much better than even high quality concrete cover, but only preformed sheet membranes or very thick coatings like mastic asphalt will bridge active cracks and prevent carbonation near them. If carbon dioxide penetrates at active cracks in the concrete surface, the alkalinity in such areas will eventually be lost and the reinforcement will be able to rust.

5.3 Application of coatings.

Sheet-coatings (membranes) can be applied under any conditions where satisfactory adhesion can be demonstrated by trying to peel the sheeting off. Coatings applied as liquids need the concrete to be clean and reasonably dry for the coating to stick properly and be resistant to peeling or blistering. Water-abrasive blasting or steam cleaning are the only methods found satisfactory for cleaning large areas of old concrete, but it is possible to coat clean new concrete without much cleaning provided that the surface is free from the oil or wax used to release the forms. Sprayed-on coating membranes prevent adhesion if they have a waxy basis, but some resin-based curing membranes improve the adhesion of coatings which are applied subsequently.

(See references 9, 11, 12, 16, 19, 24, 26, 28, 29, 33 for further information on topics in this section).

Waterproof treatments to reduce the penetration of salt water into new concrete – general information

Type of coating	Protective material	Efficiency	Life
1. Hot-applied mastic asphalt tanking (black)	Bitumen	Very high if the coating complete. Able to span cracks.	Over 50 years if covered by backfill. Probably 20 years or more if exposed, depending on climate.
2. Tanking with heavy preformed sheet materials	1. Bitumen on inorganic fabric base alone or in combination with other materials (black) 2. pvc, polyurethane, butyl rubber (black), chlor-sulphonated polyethylene, polychloroprene	Very high if properly bonded and if coating is complete. Able to span cracks.	Over 20 years if covered by backfill. Up to 20 years if exposed, depending on exposure. Between 5 and 20 or more years depending on material and exposure
3. Building paper	Bitumen on organic paper	Can be high but only if properly bonded to the concrete so that protection is complete. Able to span cracks.	Short
4. Polythene sheet	Low-density polyethylene		10 years or more if not exposed to UV light. Very short if exposed to UV (unless black).
5. Liquid surface coatings	Pitch or coal-tar epoxy with 250 μm dry coating thickness (black)	Very high if well applied and free from pinholes. Not able to span active cracks. Should be applied in more than one coat.	Up to 20 years, depending on exposure.
	Epoxy resin, two-pack polyurethane coatings 250 μm thick	Very good, but recoating is difficult because of poor inter-coat adhesion. Not able to span active cracks. Should be applied in two coats. Second coat of contrasting colour while first is still tacky.	Depends on thickness and conditions. Thick (250 μm) coatings can have long life of more than 10 years.
	Solvent-based acrylic, methacrylate, styrene-acrylic, one-pack polyurethane, chlorinated rubber	Good: these paints combine protective and decorative qualities. Multi-coat system essential to reduce incidence of pinholes. Not able to span active cracks.	Up to 10 years, depending on exposure

Type of coating	Protective material	Efficiency	Life
5. Liquid surface coatings (*continued*)	Emulsion-based acrylic or styrene-butadiene polymers and co-polymers with or without other materials.	Fair: emulsion-based paints are not truly impermeable, but still give some resistance to water penetration. Some emulsion paints are as effective as solvent-based paints in protecting against CO_2 corrosion. Not normally able to span active cracks.	Up to 10 years, depending on exposure.
6. Water-repellent treatments	Silicone, silane, siloxane	These coatings make the surface water-repellent without making it waterproof. These materials are easy to re-apply after the years and some form good primers for other types of coating. Concrete must be dry to ensure good penetration.	Probably about 10 years, but can be longer.

6 Cathodic protection

Reinforcement in concrete can rust if it is not protected by the alkalinity of the concrete, or if salt contamination has destroyed that protection. The rate of rusting may be so slow that it is insignificant, or it may be fast enough to produce serious damage in a very few years. Local variations in factors like moisture content, oxygen availability, permeability and the amount of contamination, control the rate of rusting because they determine the electrochemical conditions which cause or allow the corrosion currents to flow.

It is possible to stop rusting altogether by passing a small d.c. current to the reinforcement in such a way that no part of the reinforcement can reach an electric potential that would allow it to rust. In this way every part of the reinforcement is artificially made slightly cathodic to an externally applied anode at or near the concrete surface and electrons from the imposed protection current are supplied to all parts of the reinforcement, whereas they would flow from it in anodic areas that were corroding.

Where a structure is extensively contaminated with chlorides, cathodic protection may offer an alternative to conventional treatments because parts of the structure where contaminated concrete is still undamaged do not have to be cut away. In conventional repair, all chloride contaminated concrete has to be cut away where it is adjacent to reinforcement because corrosion will start in these areas as soon as the damaged areas are repaired. This happens because the corrosion taking place at the adjacent damaged areas provides local cathodic protection for adjacent areas of contaminated concrete. When cathodic protection is to be applied, only the damaged concrete needs to be cut out and replaced.

The comparative economics of cathodic protection and conventional treatment depend on the circumstances. Where access is easy and a maintenance team is kept on hand, patch-repair of problem areas as they show up may be quite economical. Where access is difficult and a repair needs to be trouble free for a long period, cathodic protection may be cheaper than the alternative of removing and replacing all the contaminated concrete adjacent to the reinforcement.

Although cathodic protection of concrete has been in use for about 20 years on highway bridge decks in USA, and for shorter periods on increasing numbers of other types of structures in USA and elsewhere, when applied to buildings and structures, cathodic protection is still to some extent a pioneering technique with problems of application that have to be solved job by job.

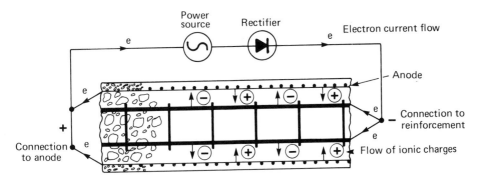

Current and ionic charge flow in cathodic protection.

One of the main problems common to all applications is the need to install durable permanent anodes in electrical contact with the concrete surface, and the need to design a system which will supply the required current in spite of the very large local variations in resistance between the surface of the concrete and the various layers of reinforcement.

Anodes can be of various forms and be made from a variety of materials; examples are conductive paints based on graphite and used for buildings and soffits, mesh made from corrosion-resistant wire, sometimes plated with precious metal or sheathed with conductive polymer and protected by a cementitious overlay, and discrete anodes in the form of plates or wires which can be fixed to or embedded in the concrete surface.

The design of a cathodic protection system must allow for the range of conditions at the reinforcement caused by variations in cover, moisture content, resistivity and oxygen availability. To do this requires careful examination of the structure to identify areas of internal splitting, checking of the electrical continuity of the reinforcement, the electrode potentials and the depth of cover. The protection system must then be divided into separate areas to distribute the current correctly to the reinforcement.

Once installed, cathodic protection systems need to be monitored to check that the protection is at a suitable level in different parts of the structure. This can be done with embedded half-cells and the automatic recording of potential measurements which can be compared with measurements made before the system was installed. Adjustments are then made to the applied current in the light of the changing circumstances.

'Overprotection' of some areas, which generates hydrogen as a result of electrolysis, cannot altogether be avoided in practice and some hydrogen evolution has to be accepted. Good control in a well-designed system keeps this to a minimum and its presence does not harm.

Beware, however, of the fact that this reassurance does not apply to prestressed concrete. Cathodic protection must not be used on prestressed concrete because there is a danger that the steel will become embrittled by the presence of hydrogen.

In spite of the formidable problems of applying cathodic protection to reinforced concrete, the technique is developing rapidly and has already been used to protect major structures in several parts of the world. The experience that is being obtained has not yet been widely disseminated or evaluated, but suggests that cathodic protection may be an economical method of preventing deterioration in certain types of structure if they are at risk as a result of chloride contamination, but are not yet extensively damaged by corrosion.

(See references 7, 9 to 12, 16, 21, 22, 24, 28, 32 for further information on topics in this Section)

Applying sprayed concrete to protect an anode installation.

References

1. *The durability of steel in concrete: Part 1. Mechanism of protection and corrosion.* Digest 263. Building Research Establishment, Garston, Watford, 1982
2. *The durability of steel in concrete: Part 2. Diagnosis and assessment of corrosion-cracked concrete.* Digest 264. Building Research Establishment, Garston, Watford, 1982
3. SOMERVILLE G. The design life of concrete structures. *Structural Engineer,* **64A**, Feb 1986, 60–71
4. PATERSON A.C. The structural engineer in context. *Structural Engineer,* **62A**, No 11, Nov. 1984
5. SLATER J.E. *Corrosion of metals in association with concrete.* ASTM Special Publication 818, Philadelphia, 1983
6. BRITISH STANDARDS INSTITUTION *BS 8110 Structural use of concrete: Part 1. Code of practice for design and construction.* BSI, London, 1985
7. TUUTTI K. *Corrosion of steel in concrete.* Research Report 4/82. Swedish Cement and Concrete Research Institute, Stockholm, 1982
8. FOOKES P.G., KAY E.A. and POLLOCK D.J. *Concrete in the Middle East.* Eyre and Spottiswood, London, 1982
9. WILKINS N.J.M. and LAWRENCE P.F. *Fundamental mechanisms of corrosion of steel reinforcement in concrete immersed in seawater.* Concrete in the Oceans Technical Report No 6, Cement and Concrete Association, Wexham, Slough, 1980
10. PAGE C.L. and TREADAWAY K.W.J. Aspects of electrochemistry of steel in concrete. *Nature,* **V**, 279, London, 1982
11. *Coordinating report. Concrete in the Oceans Programme.* Department of Energy, CIRIA, London, 1986
12. ALLEN R.T.L. and EDWARDS S.C. *Repair of concrete structures.* Blackie, London and Glasgow, 1987
13. TURTON C.D. *Plastic cracking of concrete.* Construction Guide. Cement and Concrete Association, Wexham, Slough, 1981.
14. *Simplified method for the detection and determination of chloride in hardened concrete.* IS 12/77 Building Research Establishment, Garston, Watford, 1977
15. LEES T.P. *Field tests for chloride in concreting aggregates.* Construction Guide. Cement and Concrete Association, Wexham, Slough, 1982
16. *Repair of concrete damaged by reinforcement corrosion.* Technical Report No 26. Concrete Society, London, 1984
17. HARRISON T.A. *Early age thermal crack control in concrete.* Report 91, CIRIA, London, 1981
18. *The investigation and repair of concrete highway structures.* Departmental Advice Note BA 23/86, Department of Transport, London, 1986
19. ROBERTS M.H. *Carbonation of concrete made with dense natural aggregates.* IP 6/81, Building Research Establishment, Garston, Watford, 1981
20. STILLWELL J.A. *Exposure tests on concrete for offshore structures.* Concrete in the Oceans Technical Report No 8, Cement and Concrete Association, Wexham, Slough, 1983
21. *Test method for half-cell potentials of reinforcing steel in concrete.* Standard C 876-80. ASTM, Philadelphia, 1983
22. VASSIE, P.R.W. Reinforcement corrosion and the durability of concrete bridges. Proceedings Institution of Civil Engineers, **76**, Pt. 1, Aug 84, 713–723
23. BRITISH STANDARDS INSTITUTION
    BS 1881:
    Part 5: 1970: Methods for testing hardened concrete for other than strength.
    Part 6: 1971: Analysis of hardened concrete.
    Part 114: 1983: Methods for determination of density of hardened concrete.
    Part 120: 1983: Method for determination of the compressive strength of concrete cores
    Part 122: 1983: Method for determination of water absorption.
    Part 201: 1986: Guide to the use of non-destructive methods of test for hardened concrete.
    Part 202: 1986: Recommendations for surface hardness testing by rebound hammer.
    Part 203: 1986: Recommendations for measurement of velocity of ultrasonic pulses in concrete.

24. MANNING D.G. and BYE D.H. *Bridge deck rehabilitation Manual, Part 1: Condition Surveys, Part 2: Contract preparation.* Ontario Ministry of Transportation and Communications, Research and Development Branch, Ontario 1983, 1984

25. *The durability of steel in concrete: Part 3 The Repair of reinforced concrete.* Digest 265, Building Research Establishment, Garston, Watford, 1982

26. *A guide for the use of epoxide resins with concrete for building and civil engineering.* Advisory Note 12. Cement and Concrete Association, Wexham, Slough, 1968

27. *The CIRIA guide to concrete construction in the Gulf region.* Special Publication 31, CIRIA, London 1984

28. *Materials for the repair of concrete highway structures,* Departmental Advice Note BD 27/86. Department of Transport, London, 1986

29. *Safety precautions for users of epoxy and polyester resin formulations,* Federation of Resin Formulators and Applicators, Aldershot, 1981

30. CALDER A.J.J. *Repair of cracked reinforced concrete: Assessment of injection methods.* Research Report 81. Transport and Road Research Laboratory, Crowthorne, 1986

31. WILLS J. *Epoxy coated reinforcement in bridge decks.* TRRL, Special Report 667, 1982, 23 pp.

32. KENDELL K. and LEWIS D.A. *Bridge decks; cathodic protection.* Contractor Report 4. Transport and Road Research Laboratory, Crowthorne, 1984

33. *Protection of reinforced concrete by surface treatments.* Technical Note 130, CIRIA, London 1987

Appendix: a review of the research and recommendations regarding chloride associated reinforcement corrosion in the UK and USA

Many older structures which have given a very long and creditable service life are now, not surprisingly, in need of repair. In addition, a few other structures built in the 60's and early 70's are showing signs of deterioration, though, in many cases they have not lasted their expected service life. In most cases, the primary cause of this early deterioration can be linked to chlorides in the concrete causing the reinforcement to rust, damaging both the structural integrity and appearance of the building.

Chloride salts are present to some extent in most, if not all structural concrete, and may be derived from a number of sources, the most common of which are probably as follows:

1. Calcium chloride admixtures
2. Sea dredged aggregate
3. Exposure to marine conditions
4. Chemical deicing salts

The amount of chloride derived from such sources may be considerable.

The uptake of chloride from marine sources and deicing salts is highly variable, but amounts of up to 2.5% have been determined in the surface layers of concrete bridge decks. In North America the use of deicing salts on highways and bridge decks has led to chlorides penetrating the concrete and reaching the reinforcing steel leading to cracking and spalling. The use of deicing salts in Northern Europe began later, but here it has been common practice to provide a waterproof coating to concrete bridge decks and hence the amount of penetration, in comparison, has been slight. However, at bridge abutments, cross-heads and piers where surface water containing deicing salts has drained from the bridge deck problems have occurred when chloride ions have reached the reinforcement of the supporting structures.

Impurities contained in mixing water, sand and coarse aggregate also contribute to the total chloride content. The amount of chloride obtained from the mixing water may be small, a typical mix water may contain 100 mg/litre chloride ion (potable water), which with a water/cement ratio of 0.45 would give only 0.005% chloride ion by weight of cement, it should nonetheless be taken into account.

During the 1960s and very early 1970s the construction boom throughout Northern Europe and North America put considerable pressure on the construction industry to introduce innovative techniques which would improve production rates. One such technique was the widespread use of calcium chloride as an admixture in the precasting industry. This improved productivity by enabling quicker turnaround of forms and in the production of conventional reinforced concrete to accelerate setting, particularly in cold weather.

Problems with reinforcement corrosion were also severe in the Arabian Gulf region where high temperatures coupled with a highly saline environment, created ideal conditions for corrosion. Here too, pressures on the construction industry during the boom years of the 1970s and the absence of a concrete technology based on local conditions and tailored to local requirements, meant that standards of construction were often inappropriate, and consequently severe deterioration of even relatively new structures has occurred.

The presence of chlorides in concrete is now almost universally believed to be the primary cause of reinforcement corrosion. However, there is still much debate about the critical levels of chloride, particularly calcium chloride, in concrete at which depassivation occurs and corrosion can begin. This debate has a long and varied history which is summarised in the Table which follows.

Table: Review of the history of research and recommendations regarding chloride-associated reinforcement corrosion
(Except where otherwise stated all percentages are by weight of cement)

Research in the UK and Northern Europe	Official Specifications Codes of Practice in the UK	Research in the USA	Official Specifications and Codes in Practice in the USA
	1885: Millar and Nichols. Patent for the use of calcium chloride as an accelerator for Portland cement. Pat. No. 2885.	1923: Cottinger and Kendall. After one year of investigation Cottinger found no corrosion of embedded steel when 10% calcium chloride admixture had been used. 1923: Pearson. After studying reinforced concrete with 2% calcium chloride admixture over a six year period concluded "Calcium chloride had little effect on reinforcement". 1923: US Bureau of Standards "Corrosion of embedded metals in concrete containing calcium chloride was neither serious nor progressive."	
1943: Newman. The addition of 2% calcium chloride by weight of cement appeared to cause some slight initial corrosion of steel, but that this did not progress with age.			

Table continued

Research in the UK and Northern Europe	Official Specifications Codes of Practice in the UK	Research in the USA	Official Specifications and Codes in Practice in the USA
		1947: Vollmer Steel embedded in high slump concrete displayed rust at early ages but concluded corrosion was not progressive.	
	1948: CP114 Code of Practice for the Structural use of reinforced concrete in buildings. Calcium chloride may be used to accelerate the rate of hardening—usually $1\frac{1}{2}\%$ by weight of cement will prove sufficient, and there are dangers associated with excess.		
1950: Scott, Glanville and Thomas. Calcium chloride useful in cold weather, but an excessive amount could cause setting to be too rapid and may result in corrosion of reinforcement. 2% calcium chloride by weight of cement should not be exceeded.			
1951: Institution of Structural Engineers. When an accelerator such as calcium chloride is used in a mix, care should be taken to avoid the risk of initial set occurring inconveniently soon.			

1952 to 1960
Repeated failures of
oil—tempered prestressing
strand in structures in Northern
Europe.

1952: Shideler.
Referred to studies undertaken by
the Pennsylvania Department of
Highways, which showed that steel
bars immersed in a 2% calcium
chloride solution showed less
corrosion than similar bars
immersed in tap or distilled water,
and concluded that calcium chloride
was not detrimental to embedded
metal.

1952
Failure of the Regina pre-stressed
concrete pipeline, Saskatchewan.
Calcium chloride attributed as the
primary cause of breakdown.

1954: ACI Committee 212.
    Admixtures for concrete.
Suggested no limits on the
quantity of calcium chloride
admixture which could be added
to reinforced concrete. It stated
that "calcium chloride does not
cause corrosion of embedded
steel.

1954: Veits.
Observed severe corrosion of
prestressing wires in test beams
made with 2% calcium chloride
by weight of cement, steam
cured and stored dry for $2\frac{1}{2}$
years. Some wires had corroded
more than halfway (1.0mm)
through.

Table continued

Research in the UK and Northern Europe	Official Specifications Codes of Practice in the UK	Research in the USA	Official Specifications and Codes in Practice in the USA
	1957: Code of Practice for the Structural use of reinforced concrete. Modified to read "Calcium chloride may be used to accelerate the rate of hardening of Portland cement concrete but not more than 2% be weight of cement should be used.	1957: Evans. Investigated the effects of adding calcium chloride to concrete during the manufacture of prestressed specimens. This study was modelled after the manufacture of the units used in the Regina pipeline. A number of important observations were made: i) Corrosion was not stress corrosion but galvanic corrosion. ii) Steam curing of calcium chloride admixed concrete always produced corrosion. iii) Curing with hot water did not produce such severe corrosion. iv) Natural or atmospheric curing produced the lowest amount of corrosion observed. v) Neither the magnitude of stress nor the method of manufacture had any effect on the observed corrosion.	
	1959: CP115. Code of Practice for Prestressed Concrete. "Calcium chloride should not be used when steam curing is employed". "Until more is known regarding corrosion the use of calcium chloride or salt cannot be recommended. There maybe dangers associated with excess.		

1960: Walley and Bate.
The use of calcium chloride in concrete subjected to steam curing helps to offset the loss of strength at greater ages which would otherwise be experienced with respect to concrete cured normally.

Calcium chloride should not be used in steam cured units with pretensioned steel as it caused corrosion of the reinforcement even when used in proportions not exceeding 2% by weight of cement.
Its use in post-tensioning where the tendon is remote from the concrete maybe less objectionable but that insufficient is known of its effects to justify its use. Until more is known calcium chloride could not be recommended for concreting in cold weather.

1960: Monfore and Verbeck.
Maintained that the corrosion of some prestressing wire was stress induced. The number of failures was exceedingly small in comparison to the vast number of prestressed structures in use. From experimental work they found:

i)   The amount of water soluble calcium chloride varied considerably with the C_3A content of the cement, being reduced by 50% with 3.7% C_3A and 75% with 12.6% C_3A.

ii)   The use of 3% to 4% calcium chloride by weight of cement or more may lead to serious corrosion or rupture.

iii) The evaluation of hazards associated with the use of small quantities of calcium chloride (less than 2%) was impossible to assess over the duration of the experiment. They recommended "because of the hazards involved, calcium chloride should not be used in prestressed concrete".

1960: Godfrey.
Strands from prestressed beams with 2% calcium chloride after three years outdoor exposure displayed rusting, a 5% loss of tensile strength and a 60% loss in elongation. Similar strands from control beams with no added calcium chloride showed no loss in strength or elongation. He recommended that calcium chloride not be permitted in the manufacture of prestressed concrete.

Table continued

Research in the UK and Northern Europe	Official Specifications Codes of Practice in the UK	Research in the USA	Official Specifications and Codes in Practice in the USA
1962: Roberts. The additional corrosion produced by the use of flake calcium chloride up to 2% of cement in dense normally cured Portland cement concrete would have no structural significance. Increasing amounts of corrosion were observed with increased quantities of calcium chloride. When 5% calcium chloride was added corrosion seemed to be progressive. In all cases corrosion was greater in steam cured specimens than in normally cured specimens.			
1963: Blenkinsop. The use of 2% of flake calcium chloride by weight of cement in dense well compacted concrete will have little effect on the degree of corrosion of the reinforcement. In porous or badly compacted concrete the effect of calcium chloride is to increase the amount of corrosion which would normally have taken place.		1963: ACI Committee 212. Admixtures for concrete. The use of calcium chloride had not been found to promote corrosion of reinforcement where adequate cover was provided for the steel, but should not be used where stray currents are expected or in prestressed steel.	

1965: CP 116, Code of Practice for the structural use of precast concrete. Calcium chloride is not recommended either as an admixture or internally mixed with cement in any form of prestressed work with either pretensioned or post-tensioned steel.

It should never be used in prestressed concrete which is subjected to exposed moist, warm conditions.

The total amount of calcium chloride in conventionally reinforced concrete should not exceed 2% (1.5% anhydrous $CaCl_2$) and should be dissolved in some of the mixing water.

Many admixtures contain calcium chloride and the engineer should ensure that the quantity present does not exceed the above recommendation.

Where calcium chloride is used in concrete, not less than 25 mm of cover should be given to all steel unless permanent protection is provided.

When reinforced concrete is steam cured the addition of calcium chloride is undesirable.

Calcium chloride should not be used in concrete made with sulphate resisting cement.

Table continued

Research in the UK and Northern Europe	Official Specifications Codes of Practice in the UK	Research in the USA	Official Specifications and Codes in Practice in the USA
1966: McIntosh. Calcium chloride has been well proved in practice as an accelerator for Portland cement concrete, and is so cheap and readily available that there seems little point in using any other material for that purpose.			
		1970: National Highways Research Board. The incidence of corrosion and failure in prestressed concrete bridges is rare. On the rare occasions when serious steel corrosion had occurred, it could have been avoided by application of proper design and construction practice.	
1971: Treadaway. The durability of prestressed steel in concrete could be achieved with the use of dense, impervious and uniform concrete free from chloride and with an adequate depth of cover to the steel.		1971: ACI Committee 318, Building code requirements. Admixtures for prestressed concrete or concrete which will have aluminium embedments shall not contain deletereous amounts of chloride ion. (but did not define deletereous amounts).	

1972: Handbook on the unified code of Structural Concrete. Calcium chloride is a most useful admixture as it can be of considerable benefit particularly when correctly used in winter building, in emergency repair work and in enabling a quick turnaround of formwork. It is liable to lead to some detrimental effects in the hardened concrete particularly if mis-used.

1972: CP 110, Code of Practice for the structural use of concrete. Marine aggregates may be used provided that chloride content expressed as anhydrous calcium chloride by weight of cement does not exceed 1.0%, but where the proportion exceeds 0.1%, by weight of cement, marine aggregates must not be used with high alumina cement, in prestressed concrete, or in circumstances where calcium chloride admixtures are not permitted.

In concrete containing embedded metal calcium chloride must not be added in such proportion that the total from the admixture and the total calculated from the aggregates exceeds 1.5% by weight of cement.

For concrete with embedded metal the anydrous calcium chloride should never exceed 1.5% and therefore extra-rapid hardening cements should never be used.

Calcium chloride should never be used in prestressed concrete nor in the main concrete of post-tensioned prestressed concrete unless there is an impermeable and durable barrier in addition to any grout between the main concrete and the tendons.

Calcium chloride should never be used with high alumina cement, sulphate resisting cement or supersulphated cement.

Table continued

Research in the UK and Northern Europe	Official Specifications Codes of Practice in the UK	Research in the USA	Official Specifications and Codes in Practice in the USA
	(1972:CP110)		
	Whenever calcium chloride is included in concrete containing embedded metal there is an increased risk of corrosion. It is therefore important to ensure the calcium chloride is thoroughly mixed to minimise the variations in chloride concentrations. The corrosion risk is further increased when concrete containing added chlorides is cured at elevated temperatures or subsequently exposed to warm environments.		
1974 Following the collapse of a twelve-year old post-tensioned beam and other corrosion problems in reinforced concrete the Property Services Agency prohibit its use.		1974: Clear. Proposed the concept of threshold corrosion limits in determing repair or replacement strategies by measuring the chloride content of core samples. The corrosion threshold limit for chloride was defined as "the minimum quantity of total chloride required to initiate rebar corrosion when sufficient water, oxygen and other necessary factors are present". The suggested limits were:	1974: ACI Committee 318. Building code requirements. Provision was added concerning the chloride ion content of mixing water (including that contributed as free moisture in the aggregate), to be used in prestressed concrete or concrete with aluminium embedment. It suggested that chloride ion contents greater than 400 or 500 ppm might be considered dangerous.

● Less than 1.0 lb chloride ions per cubic yard (0.6 kg/m³)—leave the concrete intact (approximately 0.2% by weight of cement).
● greater than 2.0 lb chloride ions per cubic yard (0.6 kg/m³)—remove the concrete below the top mat of bars or replace the entire deck (aaproximately 0.4% by weight of cement).
● between 1.0 lb ans 2.0 lb chloride ions per cubic yard (0.6—1.2 kg/m³), the questionable area, the decision whether or not to remove the concrete depends on the engineers willingness to accept the risk and potential cost of future corrosion problems.

1975: Institution of Structural Engineers.
Informed the BSI that in their view calcium chloride should not be used in structural concrete.

1976: Rosa.
Termed the confusion arising over the use of calcium chloride as an admixture and as a deicing salt "the chloride psychosis".

Table continued

Research in the UK and Northern Europe	Official Specifications Codes of Practice in the UK	Research in the USA	Official Specifications and Codes in Practice in the USA
	1977: CP 110, Code of Practice for the structural use of concrete admendments. Calculations on the background content of chloride which could be expected from natural sources would be less than 0.06% by weight of cement. It also sets limits for the total allowable chloride content of various types of concrete:	1977: Cook and McCoy. The chloride threshold limits are often misapplied, leading to unduly strict specifications, limiting chloride content to that suggested for bridge decks. The threshold limit was set by studies on spalled and cracked concrete caused by the external application of deicing salt, which form an entirely different environment to the service conditions of a concrete building.	977: ACI Committee 201, Guide to durable concrete in agreement with ACI Committee 222. Corrosion of metals in concrete. Table of allowable chloride contents for corrosion protection in various service environments.
	Type of use of concrete		*Type of concrete and service conditions encountered:*
	Maximum total chloride content expressed as percentage of chloride ion by weight of cement.		Maximum total chloride content expressed as a percentage of chloride ion by weight of cement.
	Prestressed Concrete. Structural concrete that is steam cured. Concrete for any use made with cement to BS4027 or BS248:		*Prestressed concrete:*
	0.06		0.06
	Reinforced concrete made with cement complying with BS12. Plain concrete made with cement complying with BS12 and containing embedded metal:		*Conventionally reinforced concrete in most environments exposed to chlorides:*
	0.35 for 95% of test result with no results greater than 0.50.		0.10.
			Conventionally reinforced concrete in moist environments not exposed to chlorides:
			0.15.
			Concrete for above ground buildings where the concrete would stay dry:
			No limit set.

1980: Everett and Treadaway. Suggested limits for the chloride content of concrete in existing buildings for the assessment of its present condition and future performance.

Low: less than 0.4%
Medium: 0.4% to 1.0%
High: Greater than 1.0%.

They stated there was evidence to show that dense well compacted concrete containing low levels of chloride in general provided good protection to reinforcement over long periods of time.

The risk of corrosion increased in concretes containing medium levels of chlorides and in concretes containing high levels there was a high risk of corrosion. It was also noted that the levels of risk are increased in all cases if significant carbonation of the concrete has occurred.

The report also stated "research has shown that the threshold value for chloride content in concrete necessary for corrosion of embedded steel can be as low as 0.15% by weight of cement."

Table continued

Research in the UK and Northern Europe	Official Specifications Codes of Practice in the UK	Research in the USA	Official Specifications and Codes in Practice in the USA
			1983: ACI Committee 318. Building Code requirements. After strong representations from contractors and admixture manufacturers, more relaxed limits were adopted than those proposed by ACI Committee 201. *Type of concrete and conditions of service:* Maximum total chloride content expressed as percentage of chloride ion by weight of cement. *Prestressed concrete:* 0.06. *Conventionally reinforced concrete in a moist environment exposed to chlorides:* 0.15. *Conventionally reinforced concrete in a moist environment not exposed to chlorides:* 0.30.

1984: Stark.
Corrosion problems reported from a survey of prestressed concrete manufacturers, apparently resulted from exposure to chlorides in the service environment, particularly deicing solutions. Corrosion also occurred after 2 to 3 years in concrete with or without calcium chloride admixture. From experimental work he observed: The threshold limit for water soluble chloride above which corrosion of prestressing tendons occurred during and immediately after curing was between 0.11% and 0.17% by weight of cement.

The threshold limit for water soluble chloride above which corrosion occurs at later ages is dependent on environmental factors including lack of dissolved oxygen, high electrical resistivity and a uniform environment along individual tendons. Lower water/cement ratios, as they influence moisture, chloride and oxygen diffusion through the concrete provide greater corrosion protection under differential or cyclic exposure conditions.

Table continued

Research in the UK and Northern Europe	Official Specifications Codes of Practice in the UK	Research in the USA	Official Specifications and Codes in Practice in the USA
	1985: BS8110, British Standard for the structural use of concrete. "Calcium chloride and chloride-based admixtures should never be added to reinforced concrete, prestressed concrete and concrete containing embedded metal". It also contains a table of total chloride content of concrete from all sources:	1985: Anderson and Black. From observations of ten years of use of calcium chloride in concrete structures, 1951 to 1961, several years of a laboratory experiments and a careful review of the literature, that in conjunction with impervious well cured concrete, it is safe to use calcium chloride in small quantities (about 2%).	
	Type of concrete: Maximum total chloride content expressed as a percentage of chloride ion by mass of cement (inclusive of pfa or ggbfs when used).	They agree that under certain environmental conditions galvanic corrosion will occur, but it requires adequate amounts of moisture, oxygen and electrolyte at the steel concrete interface. But since precast, prestressed concrete is generally produced with low water/cement ratios, is well consolidated and well cured these conditions rarely, if ever, arise.	
	Prestressed concrete, Heat-cured concrete containing embedded metal: 0.10.	They recommend a review of the present limits and consider any revisions should reflect experience.	
	Concrete made with cement complying to BS4027 or BS4348: 0.20.		
	Concrete containing embedded metal made with cement complying with BS12, BS146, BS4246 or combinations with ggbfs or pfa: 0.40.		

1986: Department of Transport Departmental advice note BA 23/86.

Chloride ions from deicing salts are the primary cause of reinforcement corrosion in highway structures.

In all cases where concrete contains more than 0.3% chloride ion by weight of cement at the position of the reinforcement, the steel is vulnerable to corrosion.

1986: Department of Transport Highways and Traffic Department Standard BD 27/86.

Reviewing materials specifications for the repair of concrete highway structures "The total chloride ion content of the materials shall not exceed 0.3% of the mass of cement. Any chloride or admixtures containing chloride salts shall not be used."

Index